Dissertation for PhD Thesis

Metabolic and bioprocess engineering of production cell lines for recombinant protein production

by Betina da Silva Ribeiro

January 2011

Bibliografische Information der Deutschen Nationalbibliothek

Die Deutsche Nationalbibliothek verzeichnet diese Publikation in der Deutschen Nationalbibliografie; detaillierte bibliografische Daten sind im Internet über http://dnb.d-nb.de abrufbar.

ISBN 978-3-8325-3123-2

Logos Verlag Berlin GmbH
Comeniushof, Gubener Str. 47,
10243 Berlin
Tel.: +49 (0)30 42 85 10 90
Fax: +49 (0)30 42 85 10 92
INTERNET: http://www.logos-verlag.de

I hereby declare that this submission is my own work and that, to the best of my knowledge and belief, it contains no material previously published or written by another person nor material which to a substantial extent has been accepted for the award of any other degree or diploma of the university or other institute of higher learning, except where due acknowledgment has been made in the text.

Betina da Silva Ribeiro

Bielefeld, 2011

To my mother, Maria da Conceição, father, António,
sister, Daniela, brother, Cristiano,
and nieces, Mariana and Marta.
To all my family and friends.
Always present in my heart.

Abstract

The small non polar CO_2 molecule challenges the industrial mammalian cell culture technology in the way that it has become a key process parameter for process design. The dissolved CO_2 changes form in aqueous solutions and affects directly the intracellular pH (pH_i) value as it does so influencing therefore important cellular processes. Nowadays, only few industrial relevant accurate data are available which make this variable very difficult to investigate. The enzyme carbonic anhydrase II (CAII) catalyzes the equilibrium of CO_2 in aqueous solutions and because it alters the speed at which this equilibrium is reached it was identified as a strong candidate for metabolic engineering. A stable hCAII expressing cell line was generated and characterized permitting the investigation of the CO_2 effects in simulated CO_2 acid load and high CO_2 levels that may occur on mammalian cell culture. A all process design for characterizing the influence of high CO_2 in bioreactor environment was employed and a proteomic analysis was performed to identify potential candidates involved in CO_2 metabolism of cells to overcome high CO_2 levels. The cell line expressing active hCAII enzyme presented a better initial re-alkalinization of cytoplasm after induced CO_2 acid load. Additionally, this result suggested that pH_i fine tuning was performed by the Cl^-/HCO_3^- exchanger (AE) and Na^--dependent Cl^-/HCO_3^- exchanger (NCBE) instead of the Na^+/H^+ exchanger (NHE1). Moreover, the most alkaline pH_i value associated to the lowest pH_i variations was observed for that cell line on long term increased CO_2 levels. In general, the increased CO_2 profile triggered the quicker progress of G0G1 cell cycle phase for both transfected and control cell lines. Nor the clone screening, hCAII expression or CO_2 profile had strong negative impact on product glycosylation. Notwithstanding, a significant difference on protein expression profile was observed for the hCAII expressing cell line as a result of an intensive proteomic analysis. Nevertheless, a biologic system is available that permits to perform further investigations in the crucial role of CAII in low mixing times and hydrostatic pressures in industrial mammalian cell culture.

Key-words: carbon dioxide, metabolic engineering, human carbonic anhydrase, intracellular pH measurement, recombinant proteins, bicarbonate, proteomic analysis, glycosylation.

ACKNOWLEDGEMENTS

Firstly, with distinction, I would like to thank Prof. Dr. Thomas Noll for giving me the opportunity of doing my dissertation at the Cell Culture Technology Group of the University of Bielefeld (Germany), the interesting selection of topics, its good mentoring, the motivation, the numerous suggestions and constant willingness to debate through which he has contributed substantially to this work.

To Dr. Heino Büntemeyer for the wise scientific and laboratory practice support, for the laboratory corridor talks and all the suggestions given for the execution of this project.

The company Roche Diagnostics GmbH, I thank for the provision of material and financial resources that made possible the realization of this dissertation. In particular, I thank Dr. Gabelsberger and Dr. Markus Emmler.

I thank to Prof. Dr. Kaltschmidt of the Cell Biology Department who have contributed through constant assistance and many suggestions for the molecular biology area.

To Prof. Dr. Hermann Ragg of the Cellular Genetics and his team, I thank for the good working conditions and the provision of equipment for the realization of the molecular genetic part of this work.

To Prof. Dr. Erwin Flaschel of Fermentation Technology Group and his team, I thank for the scientific support and for the availability of devices for the realization of this work.

My special thanks to Anja Mulholland, for her dedication to her diploma work, for great team work and for the friendship. Wish you all the best for your future!

To Dr. Marc Wingens, Dr. Raimund Hoffrogge and Nadine Wehmeier for all support given for the execution of the proteomic analysis part of this work. I learned a lot and it was a pleasure to work together with you.

Many thanks to Prof. Dr. Silverman and Dr. Tu of the Department of Pharmacology and

Biochemistry of University of Florida for the carbonic anhydrase activity measurements done.

To Tim Beckamnn, I thank you for conducting experiments for the quantification of the carbonic anhydrase enzyme activity during your Project work.

To Dr. Alexander Jockwer and the all Cell Culture Technology team from Research Center Jülich (Germany) for first putting me in contact with the mammalian cell culture technology world and for the excellent working atmosphere.

To Stefan Northoff, I thank for all support that you gave me during my thesis and for the friendship.

To Martin Volmer, I thank for the critical reading of this manuscript.

To all colleagues and staff from the Cell Culture Technology Group of the University of Bielefeld with whom I've established a great network, for the good cooperation, the pleasant working atmosphere and the numerous supports in many experiments.

To Holger Talinski, you were always there for me and you have strengthened me in so many difficult times. Thank you that you very much.

To all my friends in Portugal for understanding for the prolonged absences and for the emotional support. I'm glad to have you in my life!

To my Portuguese friends in Basel for their patience in accepting a lot of "No, this weekend I have no time because I have to continue writing my thesis". You are very important in life!

To all others who were not mentioned but always gave me support.

I thank my parents, sister and brother, for the love, courage and confidence you always gave through my life.

CONTENTS

1 Introduction and Setting of Tasks **1**

1.1 Introduction 1

1.2 Motivation 2

1.3 Setting of Tasks 3

2 Literature Review **5**

2.1 Large-scale production of recombinant proteins 5

2.1.1 State of the art 6

2.2 Carbonic Anhydrase 10

2.3 Intracellular pH regulation 11

3 Generation of a pCO_2 robust cell line **16**

3.1 Stable transfection of hCAII gene in CHO cells 16

3.1.1 Sequencing of hCAII gene 17

3.1.2 Transfection and selection 18

3.1.3 Verification of hCAII expression 20

3.2 Characterization of clones 21

3.2.1 Growth and metabolism 21

3.2.2 Glucose and Lactate metabolism 23

3.2.3 Ammonium metabolism 25

3.2.4 Product formation 26

3.3 Mycoplasma sterility testing with PCR 27

3.4 hCAII Activity 28

3.5 Interpretation of results 28

4 Physiologic effect of CAII on pH_i 33

4.1 Acid load 33

4.2 Discussion of the results 36

5 Effect of long-term CO_2 increase on cells 40

5.1 pCO_2-controlled cultivations 40

5.1.1 Growth 41

5.1.2 Product formation 42

5.1.3 Metabolism 43

5.2 Evolution of Intracellular pH value 45

5.3 Cell cycle distribution 46

5.4 Carbonic Anhydrase II expression during cultivation 47

5.5 Discussion 48

6 Recombinant protein glycosylation 51

6.1 Influence of hCAII-expression on glycosylation 51

6.2 Influence of long term CO_2 increase on glycosylation 52

6.3 Discussion of the results 53

7 Proteomic analysis 55

7.1 Comparison of the expression pattern of Clone E11 and Control HyQ . 55

7.1.1 Selection of suitable sampling points 55
7.1.2 Analysis of DIGE gels 56
7.2 Analysis of the effects of increased pCO_2 values on the proteome of the clone E11 59
7.2.1 Protein identification 59
7.2.2 Classification of the identified proteins into functional categories 61
7.2.3 Hierarchical cluster analysis of the identified proteins 61
7.3 Discussion of results 67

8 Conclusions **70**

9 Future perspectives **72**

10 Materials and Methods **74**
10.1 Molecular biological methods 74
10.1.1 Vectors 74
10.1.2 Strategy for vector construction 76
10.1.3 Plasmid DNA restriction 77
10.1.4 Agarose gel electrophoresis 78
10.1.5 Purification of DNA 79
10.1.6 Phenol-Chloroform extraction 79
10.1.7 Ethanol precipitation 80
10.1.8 Ligation of DNA molecules 80
10.1.9 Transformation 81
10.1.10 Plasmid isolation 82
10.1.11 Determination of DNA concentration and purity 82
10.2 Primer Design 83

10.2.1 Mycoplasma test with PCR 84
10.2.2 Sequencing 86
10.3 Nucleofection of mammalian cells 86
10.4 Cells 87
10.4.1 Bacteria 87
10.4.2 Mammalian Cells 87
10.5 Media and Supplements 87
10.5.1 Media for cultivation of *E. coli* strains 87
10.5.2 Media for cultivation of mammalian cells 87
10.5.3 Monod Kinetic 88
10.5.4 Establishing of Working Cell Bank (WCB) 90
10.6 Cultivation 90
10.6.1 Cultivation in disposable bioreactors 90
10.6.2 Clone screening 91
10.6.3 Bioreactor cultivations 92
10.7 *On-line* monitoring 94
10.7.1 pCO_2 sensor 94
10.8 Analytical Methods 95
10.8.1 Determination of bacteria OD 95
10.8.2 Determination of viable cell density and viability 95
10.8.3 Determination of metabolites glucose and lactate 96
10.8.4 Determination of metabolites amino acids 96
10.8.5 Determination of ammonium 97
10.8.6 Determination of Osmolality 98
10.8.7 Determination of product concentration 98

10.8.8 Glycan analysis 98
10.8.9 Dissolved gas analysis 98
10.9 Flow cytometry 99
10.9.1 Determination of transfection efficiency 100
10.9.2 Intracellular pH measurement 100
10.9.3 Re-alkalinization experiment 104
10.9.4 Cell Cycle 105
10.10 Protein biochemical methods 106
10.10.1 Protein extraction 106
10.10.2 Determination of protein concentration 109
10.10.3 Acetone precipitation 109
10.10.4 SDS-PAGE 109
10.10.5 Western blot 110
10.10.6 Carbonic anhydrase activity 111
10.11 Proteome analysis 112
10.11.1 Sample preparation 113
10.11.2 1st Dimension: Isoelectric Focusing 115
10.11.3 2nd Dimension: SDS-PAGE 116
10.11.4 Gel scanning 117
10.11.5 Gel staining methods 118
10.11.6 Image analysis 118
10.11.7 Mass spectrometry 120

11 Bibliography **123**

12 Nomenclature **138**

13 Appendix **143**

CHAPTER 1

INTRODUCTION AND SETTING OF TASKS

1.1 INTRODUCTION

In the production of recombinant proteins, not only a high expression of the target protein is desired, but also an all process design should be taken into account to achieve high cell densities, higher productivity and product quality, decrease production costs and enhanced downstream processing. The dissolved carbon dioxide concentration (dCO_2) has been identified as one of the key process parameters affecting negatively cell growth, productivity and product quality in cell culture technology for the production of recombinant proteins. Carbon dioxide is produced by the cells themselves and could accumulate in the culture media leading to critical levels, especially in large scale industrial cultivations where the hydrostatic pressure effect and mixing times are critical parameters. This non-polar molecule easily enters the cell, and there is converted to bicarbonate and protons, through carbonic anhydrase. The accumulation of these protons in cytoplasm will cause acidification, which can interfere with optimal pH of enzymes involved in growth and metabolism.

Two main aspects are involved in the negative impact of high CO_2 levels on cells. First, in large-scale mammalian cells cultivations, the cells experiment higher hydrostatic pressures (at the bottom of bioreactor), and therefore, higher pCO_2 levels than in lab-scale [MOSTAFA AND GU, 2003]. Second, the mixing times in large-scale cultivations are considered to be in order of several seconds [TRAMPER, 1995] contributing that the cells experiment different CO_2 levels in a short time and therefore need a quick mechanism to re-adjust the intracellular pH_i.

The group of Prof. Dr. Noll (IBT-2, Research Centre Juelich, Germany) recently developed novel industrial-like processes for pCO_2-controlled mammalian cell cultivations up to 10 L-scale (fed/batch, perfusion) [Jockwer, 2008]. These do not only allow the simultaneous control of pH and pO_2 but, at the same time, also investigate the influence of defined dCO_2 levels on cell physiology and productivity. This knowledge and tools were used for the execution of this project.

1.2 MOTIVATION

Several approaches had been employed to try to circumvent the adverse effects of carbon dioxide on cell culture. These include medium design without bicarbonate [MATANGUIHAN *et al.*, 2001; GOUDAR *et al.*, 2007], stripping strategies [MOSTAFA AND GU, 2003] and metabolic engineering involving the sodium/hydrogen exchanger isoform 1 (NHE1) [ABSTON AND MILLER, 1995]. Additionally, suggestions have been made for increasing the expression of other ion exporters, such as Na^+-dependent Cl^-/HCO_3^- exchanger (NCBE) or Cl^-/HCO_3^- exchanger (AE), to contribute to mitigate the inhibitory effect of elevated dCO_2 on CHO cells [ABSTON AND MILLER, 1995].

A group of researches examined the ability of the carbonic anhydrase II to bind to and affect the transport efficiency of the NHE1 isoform of the mammalian Na^+/H^+ exchanger [LI *et al.*, 2002]. Coimmunoprecipitation tests suggested that the proteins form a complex *in vivo*. The H^+ transport rates were determined after transient induction of acid load by shifting cells from O_2-gassed nominally CO_2-free medium to CO_2/HCO_3^--containing medium. A higher initial rate of intracellular pH (pH_i) recovery was observed for the CHO cell line overexpressing both NHE1 and hCAII exchanger in comparison with a cell line overexpression only the NHE1 exchanger. The author suggested that the carbonic anhydrase II promoted the local dissociation of CO_2 to HCO_2^- and to H^+ and NHE1 directly removed the protons [LI *et al.*, 2002].

The overexpression of NHE1 alone in CHO cell lines was used to test responses to high pCO_2 [ABSTON AND MILLER, 2005]. But the inhibitory effect on cell growth at high pCO_2 and osmolality was not affected by the NHE1 level. The results of this work suggested that the hNHE1 levels in wild-type CHO cells appear to be sufficient to protect the cells from levels of metabolic byproducts typically found in cultures.

So far, carbonic anhydrase II enzyme has been cloned and overexpressed in cells but have not been examined for protection against suboptimal levels of carbon dioxide. This project gives an overview of such an investigation with industrial-like relevant approaches. The open questions are:

- how cells respond to rapid changes of pCO_2?
- does hCAII overexpression will enable the cells to withstand suboptimal levels of pCO_2 as they occur in large-scale cultivations?
- does hCAII overexpression improves the cell's re-adaptation to unphysiological pCO_2 levels?
- is the overall cell's growth, metabolism, productivity and product quality improved?

1.3 SETTING OF TASKS

This project aims the development and characterization of a production cell line based on Chinese Hamster Ovary (CHO) cells by introducing a recombinant enzyme relevant for CO_2 metabolism, the human carbonic anhydrase II. It is divided into 5 parts described below.

Part 1: The first part of this project, involved the generation of a stable production cell line expressing the human carbonic anhydrase II (hCAII). The host cell line type CHO, commonly used in cell culture technology, was used. The biopharmaceutical protein of interest is a relevant industrial recombinant protein. The selected stable clones were characterized in batch culture in comparison with the control cell line. The decision for one specific clone was made based not only on growth, productivity and metabolic data, but also on the expression and enzyme kinetics of the hCAII enzyme.

Part 2: The hCAII genetic engineered cells were exposed to weak acid in form of CO_2. These acid-load experiments were used to observe the effect of rapid changes of carbon dioxide on cell intracellular pH (pH_i) recovery rate in comparison to the control cell line.

Part 3: In this Chapter is described the characterization of the cell line generated in Part 1 and the control cell line for their behavior in industrial-like cultivations under different pCO_2 profiles in order to gain some insight and rational understanding of cellular mechanisms induced by slow increase of dCO_2 during the cultivation. This included cell growth, productivity, metabolism as well as cell physiology (cell cycle analysis and pH_i measurements).

Part 4: The influence of the hCAII expression on the product glycosylation is presented in this chapter. Product glycosylation analysis were performed for the different clones cultivated in batch mode in shaking flaks (Chapter 3) as well as for the batch bioreactor experiments (Chapter 5).

Part 5: Differential proteomic analysis were used to identify proteins being differently expressed under increased pCO_2 levels for the bioreactor cultures done in Part 3.

CHAPTER 2

LITERATURE REVIEW

Recombinant glycoprotein therapeutics have proven to be invaluable pharmaceuticals for the treatment of various diseases. Chinese hamster ovary (CHO) cells are widely used in industry for the production of these proteins. Several strategies for engineering CHO cells for improved protein production have been tried with considerable results. The focus has mainly been to increase the specific productivity and to extend the culture longevity by preventing programmed cell death [MOHAN *et al.*, 2008].

2.1 LARGE-SCALE PRODUCTION OF RECOMBINANT PROTEINS

The commercial success of monoclonal antibodies and recombinant proteins has led to the need for the very large-scale production in mammalian cell culture. This has resulted in rapid expansion of global manufacturing capacity, an increase in size of reactors (up to 20,000 L; MERTEN, 2006) and a greatly increased effort to improve process efficiency with concomitant manufacturing reduction [BIRCH AND RACHER, 2006]. Antibody titers at an industrial scale of 5 $\frac{g}{L}$ and more have already been achieved [BIRCH, 2005]. Indeed, not all scale-up issues, between others, mass transfer issues (O_2, CO_2), homogeneity, mixing time, or shear effects, have been engineered totally satisfactorily. In this work, a special focus on carbon dioxide will be made.

2.1.1 STATE OF THE ART

Mostly, oxygen demands have been considered to be the most important scale-up parameter. Oxygen is the final electron acceptor in the mitochondrial respiration chain and is directly linked to the generation of energy. In equilibrium with air the oxygen concentration is just about 0.2 $\frac{mmol}{L}$ and therefore oxygen supply to the cells is often the limiting factor, especially at high cell densities now being achieved [EIBL *et al.* 2009]. Due to the concerns related to the perceived need to enhance oxygen mass transfer by increasing the driving force and/or interfacial area by using sintered spargers, rather than volumetric mass transfer rate (k_La) by agitation or aeration rate. The problem of high levels of pCO_2 and associated high acidity plus increased osmolality associated with pH control arose and is considered to be the major difficulty encountered in scaling up mammalian cell culture to achieve high cell densities [CHU AND ROBINSON, 2001; MEIER, 2005]. The increase of driving force is performed by blending oxygen into the air or using a separate oxygen bleed into the reactor. This strategy is because to enhance k_La or air sparge rate is perceived to be more damaging to the cells and also increase the possibility of foaming. However, if the scale up is maintained at equal vvm, then the proportion of the sparged oxygen transferred to the medium and CO2 transferred into the sparged air on the small scale and the large should be the same [NIENOW, 2006].

At the initial stages of cultivation, it is necessary to control pH by the use of CO_2 sparged in the air. However, with the increasing cell density, whether from fed-batch or perfusion techniques, there is a proportionate increase in oxygen demand. Therefore, since the respiratory quotient, RQ (mol CO_2 produced per mol O_2 consumed) ranges from about 0.9 to about 1.3, the amount of CO_2 generated increases at essentially the same rate [OTZTURK, 1996; DE ZENGOTITA, *et al.*, 2002]. Thus, if k_La is maintained constant, the concentration of CO_2 in the exit air and therefore in solution increases. If a low enriched air flow rate or pure oxygen are used to enhance the driving force and thereby reduce the flow rate and potential damage due to bursting bubbles, high levels of pCO_2 of 150 - 200 mmHg are predicted to be reached [GRAY *et al.*, 1996; TATICEK *et al.*, 1998]. Such levels, have been measured in 1800-2500 L bioreactors and a high cell density perfusion bioreactor [GRAY *et al.* 1996; DE ZENGOTITA *et al.*, 2002]. Detrimental effects have been reported for growth and productivity of a recombinant myeloma process [AUNINS *et al.*, 1993], hybridoma, NSO, BHK, [DE ZENGOTITA *et al.*,

2002], CHO [GRAY *et al.*, 1996; MOSTAFA AND GU, 2003; ZHU *et al.*, 2005] and insect cells [GARNIER *et al.*, 1996] at this high levels of pCO_2, though these cells grew well in the range 35 to 80 mmHg pCO_2, comparable to the physiological range of 31-54 mmHg pCO_2 [GUYTON, 1991; DE ZENGOTITA *et al.*, 2002].

The cells continuously produce metabolic CO_2 as a waste product. The catabolism of proteins, carbohydrates, and lipids ultimately results in the formation of acetyl-CoA, which feeds into the Krebs cycle. The Krebs cycle, the primary source of energy production in mitochondria, effectively oxidizes acetyl-CoA to carbon dioxide [LEHINGER, 1982]. CO_2 is thus the primary waste product of respiratory oxidation.

The drastic effects of CO_2 in cell culture result from the fact that, as non polar molecule, the dissolved CO_2 (dCO_2) can easily across the cell membrane into the cytosol and the mitocondrial compartment affecting the intracellular pH (pH_i) and directly influencing important cellular processes [ANDERSEN AND GOOCHEE, 1994]. Inside the cell, CO_2 reacts with H_2O and forms H_2CO_3. This reaction occurs spontaneously and is also catalyzed by carbonic anhydrase enzymes (CA). Carbonic acid is a weak acid that dissociates in two steps, therefore the acid-base conversion properties of bicarbonate are: $H_2CO_3 \rightleftarrows HCO_3^- + H^+ \rightleftarrows CO_3^{2-} + H^+$, reactions governed by pKa (25°C) values of 6.4 and 10.3, respectively. This means that the substrate under investigation, the bicarbonate, changes form, and alters pH_i as it does so placing a burden on the cell to readjust the internal pH by increasing the rate of proton transporters. The situation becomes more complex because of the equilibrium CO_2 (gas) $\rightleftarrows$ CO_2 (dissolved), and it varies with partial pressure of CO_2, temperature and pH (Henry constant for CO_2, k_H (25°C) of 29.7 $\frac{atm.L}{mol}$). At physiological pH of most mammalian cells cultivations, main form is HCO_3^-. Bicarbonate is a simple single carbon molecule that plays surprisingly important roles in diverse biological processes. Among these are enzyme activity in the Krebs cycle, cellular pH, cell cycle and volume regulation. Since bicarbonate is charged it is not permeable to lipid bilayers. Mammalian membranes thus contain bicarbonate transport proteins to facilitate the specific transmembrane movement of HCO_3^- [CASEY, 2006].

Inhibition by CO_2 is therefore partly due to medium acidification if pH is not controlled. If controlled by base addition, the equilibrium will driven further to the right and thus increase osmolality, which also leads to growth inhibition [DE ZENGOTITA *et al.*, 2002]. Thus pH control strategy is very important for two reasons, one associated with temporal and spatial pH excursions associated with feed strategy; and secondly, because of its impact on medium osmolality.

But pCO_2 does not only interfere with cellular growth and productivity. It also negatively affects product quality or glycosylation of recombinant proteins [KIMURA AND MILLER, 1997; ZANGHI *et al.*, 1999]. This is also due to the intracellular pH shift which influences the activity of the glycosyltransferases and results in an inhomogeneous and modified glycosylation. As the correct glycosylation in most pharmaceutical drugs is a prerequisite for therapeutic effectiveness and a wrong glycosylation can increase immunogenicity and time plasma clearance, pCO_2 is also affecting pharmacokinetic and pharmacodynamic behavior of the drug. Therefore, wrongly glycosylated proteins have to be eliminated during downstream processing which is time consuming and cost intensive [NIENOW, 2006].

In production facilities with a large number of bioreactors, dissolved carbon dioxide sensors tend not to be used, and as a result this variable will generally go unmonitored, making the meaningful analysis of data more difficult. For aerobic fermentations, mass transfer of carbon dioxide can be described in an analogous way to oxygen transfer. The mass transfer coefficient for carbon dioxide is 0.89 times that for oxygen. The maximum dissolved carbon dioxide concentration as a function of exit gas composition is compared with the concentration obtained by assuming equilibrium between the broth and exit gas. The difference between these two concentrations is typically 20 - 40 % of the equilibrium concentration [ROYCE AND THORNHILL, 1991]. There is then a need to accurately monitor and control dCO_2 concentrations in mammalian cell culture processes. Off-line sensors are limited by the low frequency of data collection and errors introduced by sample handling and are difficult to use for continuous control. For these reasons, the use of *in situ* dCO_2 sensors is preferable. Most of these sensors utilize the Severinghaus principle, with a pH-sensitive dye and optical fiber used in place of the pH electrode which makes this probe having a short lifetime, sensitive to interfering compounds and restricted range of measurement, thus not adequate for long mammalian cell cultivations. The recent developed fiber optic dCO_2 sensors using the fluorescent dye hydroxypyrenetrisulfonate (HTPS) are designed and were successfully tested for industrial monitoring in cell culture processes [PATTINSON *et al.*, 2000].

For all reasons mentioned above there is strong interest in detailed understanding of the mechanisms of CO_2 effects in cell culture, and in the development of optimized processes and production cell lines more robust against suboptimal CO_2 concentrations.

The large-scale problem has been addressed in some depth by MEIER (2005). Using pure oxygen though sintered spargers will lead to smaller bubbles (with the greatest potential for damage when bursting) and hence locally higher driving force. Thus, they will greatly enhance the rate of oxygen transfer to the medium but at the same time, they will be very poor at stripping out CO_2 since they rapidly become saturated with it. In this case, the task of CO_2 stripping is left to the main air sparge with relatively larger bubbles (Meier 2005) for which the driving force will be less than on the small scale because of the stoichiometric ratios will be less suitable. Monitoring of pCO_2 and using a judicious combination of air, nitrogen and oxygen flow rates to control pCO_2 and dO_2 independently should allow levels of both to be achieved satisfactorily for cell culture at the 20,000 L scale [MEIER, 2005]. This approach, of using higher air flow rates, was essentially used successfully by MOSTAFA AND GU (2003) to grow CHO cells at the 1000 L scale. Interestingly, they misinterpreted their success to the use of larger bubbles sizes with a pipe sparger (at 0.01 vvm) compared to a sintered sparge stone (at 0.002 vvm) rather than to the five times higher air flow rate used with the open pipe. This increase in flow rate reduced pCO_2 from 180 to 70 mmHg. Overall, MOSTAFA AND GU (2003) reported an improved productivity (40 %), culture time (the fed-batch time was extended from 10 to 14 days) and final titer (2-fold, equivalent to that at the 1.5 L scale). However they also noted that at constant pH, higher pCO_2 is also associated with higher osmolality and that therefore the improved performance at lower pCO_2 might also be due to reduced osmolality [GARNIER *et al.*, 1996; DE ZENGOTITA *et al.*, 2002] but it was not monitored.

The larger the scale and the higher cell densities and associated oxygen demand, the greater the problems related to CO_2 concentration, acidity and osmolality will become unless higher gas flow rates can be shown not to be damaging to the cells. The successful use of an open pipe with 20 ppm anti-foam [MOSTAFA AND GU, 2003] is encouraging in this respect. However, if such an approach cannot be used, there will be a limit either to the scale of operation and/or the maximum cell density attainable at the larger scale. A question arises: does genetic engineering can overcome this problem?

2.2 CARBONIC ANHYDRASE

Carbonic anhydrase was first characterized in erythrocytes as the result of a search for a catalytic factor that would enhance the transfer of bicarbonate from the erythrocyte to the pulmonary capillaries [MELDRUM AND ROUGHTON, 1933]. Since then, the enzyme has been shown to play an important role in most acid/base-transporting [CORENA *et al.*, 2002]. The enzymes in the family of carbonic anhydrases (CA, carbonate hydro-lyase, EC 4.2.1.1) are zinc metalloenzymes that catalyze the rapid interconversion of carbon dioxide and water into carbonic acid, protons and bicarbonate ions. The catalytic mechanism can be described in two separate and distinct parts (Equation 2.1 and 2.2.

$$\text{EZn-OH}^- + CO_2 \rightleftarrows \text{EZn-HCO}_3^- \underset{H_2O}{\overset{\rightleftarrows}{\longrightarrow}} \text{EZn-H}_2\text{O} + HCO_3^- \qquad (2.1)$$

$$B + \text{EZn-H}_2\text{O} \rightleftarrows B +^{+} \text{H-EZn-OH}^- \rightleftarrows BH^+ + \text{EZn-OH}^- \qquad (2.2)$$

CAs catalyzes this reactions in either direction depending on the conditions. It does not alter the equilibrium itself, but only the speed at which it is reached. This reaction occurs spontaneously in aqueous solution, but does so slowly (hydration equilibrium constant, $k1(25°C) = 1.7x10^{-3}$). In mammals 14 carbonic anhydrase isoforms have been identified and the predominant cytoplasmic isozyme is carbonic anhydrase II (CAII) [SLY AND HU, 1995]. Carbonic anhydrase II can greatly increase the rate of the reaction, with typical catalytic rates (turnover rate) ranging between 10^4 and 10^6 reactions per second [MAREN, 1967].

Carbonic anhydrase is known to play a central role in intracellular buffering and CO_2 transport in red blood cells [MELDRUM AND ROUGHTON, 1932], in acid secretion by the stomach [DAVENPORT, 1939] and in the regulation of intracellular [ROOS AND BORON, 1981] and extracellular pH [CHEN AND CHESLER, 1992]. Certain point mutations occur without apparent clinical effect, however, complete absence leads to mild mental retardation and cerebral calcification, osteopetrosis and renal tubular acidosis [SLY *et al.*, 1985].

Carbonic anhydrase II (CAII) consists of a single polypeptide chain with 260 amino acid residues corresponding to a molecular mass of about 29 kDa present in the cytosol of most tissues, but highest concentrations are found in erythrocytes.

The Figure 2.1 presents the active site of carbonic anhydrase II showing the orientation of the proton shuttle His64 with respect to the zinc and its three ligands His94, His96 and His119. The black sphere is zinc [FISHER *et al.*, 2005].

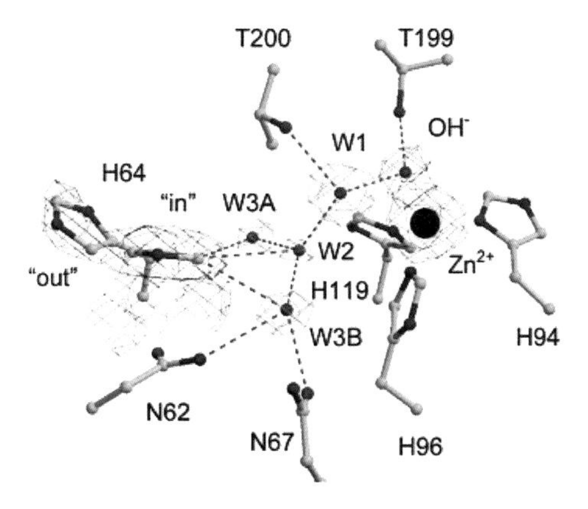

Figure 2.1: Crystal structure of the wild-type human CAII active site (pH 7.0). Water molecules and SO_4^{2-} are colored green and blue, respectively; His64 in and out conformers are colored blue, and the Zn^{2+} is colored blue. Red dashed lines connect ordered solvent and residues but do not indicate hydrogen bonds [FISHER *et al.*, 2005].

Kinetic analysis of wild-type hCAII over the pH range of 6.0 - 9.0 shows that hCAII is active over a broad pH range [KHALIFAH, 1971; STEINER *et al.* 1975]. Also, structural analysis of wild-type hCAII at pH 5.7 and 6.5 showed that the active site exhibits no major pH-dependent conformational changes, with the exception of proton shuttle residue His64 [NAIR AND CHRISTIANSON, 1991]. The position of His64 occupies a conformation oriented away from the zinc ion at pH 5.7 and is named the "out" conformation, while at pH 8.5, the side chain of His64 is in the "in" conformation, pointing toward the zinc ion.

2.3 INTRACELLULAR PH REGULATION

Intracellular pH is an important modulator of the cell function. Many enzymes exhibit pH dependence in the physiological range such their activities are affected by small variations in intracellular pH [RO AND CARSON, 2004]. Hence, precise regulation of cytosolic pH (pHi) is a prerequisite for the normal functioning of cells [FRELIN *et al.*, 1988; MADSHUS, 1988; ROOS AND BORON, 1981]. Most of cells are equipped with several mechanisms to regulate pH_i, which makes its regulation extremely complex. This becomes even more complex because several pH_i regulation mechanisms may form complex metabolon which are complex of weakly interacting binding partners passing on very efficiently substrates from one active site to another and thereby facilitating the process dramatically. The different regulation mechanisms form complex metabolons with each other through carbonic anhydrases, mainly CAII inside the cell and CAIV outside [PURKERSON AND SCHWARTZ, 2007].

There are two primary known acid extrusion mechanisms in vertebrate cells: the Na^+-H^+ exchanger (NHE), which exchanges one extracellular Na^+ for one intracellular H^+; and the Na^+-dependent HCO_3^-/Cl^- antiport (NCBE), which exchanges extracellular Na^+-HCO_3^- for intracellular HCl [ROOS AND BORON, 1981; REINERTSEN *et al.*, 1988; TONNENSSEN *et al.*, 1990; KAHN *et al.*, 1990]. Among these, the Na^+-H^+ antiport has been studied in most detail. The Figure 2.2 illustrates the major cell membrane transporters involved on the regulation of pH_i after an induced acidification of cytoplasm.

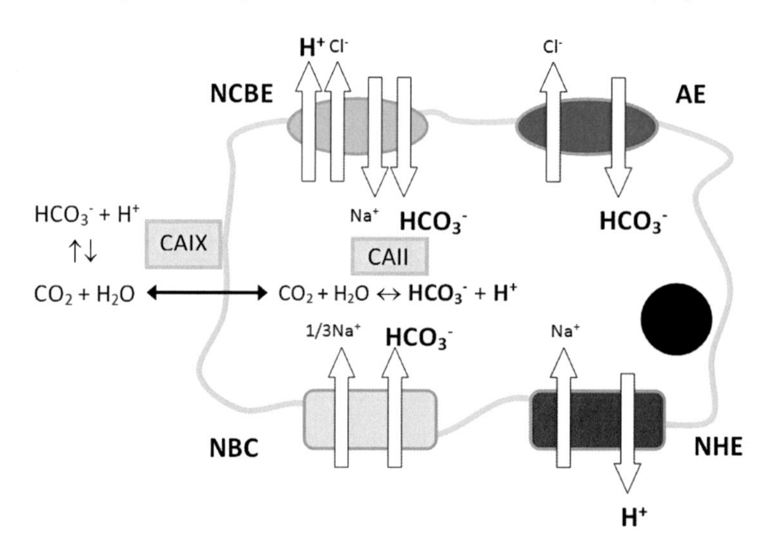

Figure 2.2: Scheme of the major cell membrane transporters involved on the regulation of pH_i after an induced acidification of cytoplasm. NBC: Na^+/HCO_3^- symporter, NCBE: sodium dependent Cl^-/HCO_3^- exchanger, NHE, Na^+/H^+ exchanger and AE, Cl^-/HCO_3^- exchanger (DOLZ *et al.*, 2004])

Mammalian Na^+/H^+ exchangers (NHE) are ubiquitous membrane ion transporters belonging to a gene family of related proteins that utilize a downhill transmembrane Na^+ gradient to energize H^+ extrusion up an electrochemical gradient in an electroneutral 1:1 stoichiometry. Therefore, it serves to regulate pH_i and cell volume, and to initiate changes in the growth or functional state of cells [ORLOWSKI AND GRINSTEIN, 1997]. To date, nine isoforms (NHE1–NHE9) have been identified within the mammalian NHE family [ORLOWSKI AND GRINSTEIN, 2004; NAKAMURA *et al.*, 2005]. The NHE1 isoform is the 'housekeeping' isoform of the exchanger, therefore has been studied in most detail, and is ubiquitously expressed in the plasma membrane of virtually all tissues [Slepkov *et al.*, 2007]. NHEs are targets for inhibition by the diuretic compound amiloride and its analogues, and by novel benzoylguanidine derivatives [COUNILLON *et al.*, 1993]. The NHE1 isoform is highly regulated. Intracellular acidosis is the major stimulus that regulates NHE1 activity, which is negligible under normal physiological conditions, but is rapidly activated as the pH_i decreases [KARMAZYN *et al.*, 2001]. The membrane domain of NHE1 is both necessary and sufficient for ion transport, whereas the cytosolic domain is involved in regulation of the activity of the exchanger [WAKABAYASHI *et al.*, 1992]. In addition to responding to intracellular protons, NHE1 is regulated by phosphorylation by various kinases and by interactions with other cellular proteins [SLEPKOV *et al.*, 2007], including calcineurin homologous protein [PANG *et al.*, 2001], calmodulin [WAKABAYASHI *et al.*, 1997], and Hsp70 [SILVA *et al.*, 1995]. One of most important cellular proteins is the carbonic anhydrase II (CAII), which binds to amino acids 790–802 at the distal end of the C-terminal tail of NHE1 creating a transport metabolon and increases the activity of the exchanger, particularly under acidic conditions [LI *et al.*, 2002]. Phosphorylation of NHE1 at a site within amino acids 634–789 causes an increased interaction between NHE1 and CAII [LI *et al.*, 2002; LI *et al.*, 2006]. Based on first cDNA, the Na^+-H^+ exchanger NHE1 has been identified in Chinese hamster [COUNILLON AND POUYSSEGUR, 1993].

The second mechanism for acid extrusion is the Na^+-dependent Cl^--HCO_3^- exchange. It was first identified in snail neurons [THOMAS, 1977] and other invertebrate preparations [ROOS AND BORON, 1981]. The SLC4A10 gene product, commonly known as NCBE for Na^+-driven Cl^--HCO_3^- exchanger, is highly expressed in rodent brain. However, some of the earlier data are not consistent with Na^+-driven Cl^--HCO_3^- exchange activity. PARKER *et al.* (2008) evidence that the Na^+-HCO_3^- cotransport activity of SLC4A10 under physiological conditions is independent of any Cl^- countertransport and thus that NCBE in fact normally functions as an electroneutral Na^+-HCO_3^- cotransporter.

Bicarbonate has a crucial role in pH homeostasis. Inside cells, it is often the most common anion, and outside it is the second most common after chloride [THOMAS, 1989]. But in practice, bicarbonate-buffered solutions are difficult to handle. For a stable pH the solution must be equilibrated with a gas mixture containing a known and fixed percentage of CO_2. Carbon dioxide readily crosses plastic, especially silicone rubber tubing and leaves solution all too readily when pumped by a peristatic pump. In many optical systems it is must easier to avoid any gas at all. Since the movement of HCO_3^- acidifies the region it leaves and alkalinizes the opposite side of the membrane, HCO_3^- transporters are clearly involved in the regulation of pH_i [CASEY, 2006].

When it was discovered that cultured mammalian cells could regulate their pH_i perfectly well with NHE1, it was widely assumed that bicarbonate would make no real difference. This assumption ignored the importance of bicarbonate in intracellular buffering [ROOS AND BORON, 1981], even without taking into account its requirement by pH_i regulators. Thus every intracellular bicarbonation can buffer four times better at any physiological pH than any ordinary buffer at its pK. For fast, buffering, carbonic anhydrase is needed. In cardiac cells, the HCO_3^- transporters are responsible for about 50 % of pH_i recovery, through alkaline HCO_3^- influx [VANDENBERG *et al.*, 1993].

Between the bicarbonate transporters, the Cl^-/HCO_3^- exchanger AE transports HCO_3^- ions from cells at alkaline pH_i to accelerate recovery from an alkaline load. The reversibility of this transporter allows transport of HCO_3^- ions into cells during acidic pH_i [LINDSEY *et al.*, 1990; ZHANG *et al.*, 1996]. The reversible Na^+/HCO_3^- cotransporter (NBC) will transport HCO_3^- ions into or out of cells [AMLAL *et al.*, 1999].

Cl^-/HCO_3^- exchange antiport was first described in cardiac muscle by VAUGHAN-JONES (1982). The activity of AE is strictly regulated by the pH_i. In vero cells, AE, which acts as an acidifying mechanism, is strongly activated at pH_i higher then 7.1 [TONNESSEN *et al.*, 1990]. CAII binds to the cytoplasmic C terminus of the erythrocyte AE1 exchanger [VINCE AND REITHMEIER, 1998], mediated by an interaction between an acidic motif on AE1 (hydrophobic residue, followed by 4 residues, with at least 2 being acidic) [VINCE AND REITHMEIER, 2000]. There are suggestions that CAII, the AE, and Na^+-H^+ exchanger activity may be linked together in a functional complex or metabolon involved in intracellular bicarbonate and pH_i regulation [YAO *et al.*, 1999; LANDESMAN-BOLLAG *et al.*, 2001].

In lymphocytes, the transporters NHE (moves H^+ out and Na^+ in) and AE (moves HCO_3^- out and Cl^- in) exchangers working together result in no change in cell pH, since the acid resulting from HCO_3^- efflux is balanced by H^+ efflux; thus, the net effect is cell loading with NaCl. Osmotic water movement restores cell volume. [MASON *et al.*, 1989].

The electrogenic Na^+/HCO_3^- co-transporter (NBC) mediates Na^+-coupled HCO_3^- transport across plasma membranes of many mammalian cells with an apparent stoichiometry of 3 HCO_3^- per 1 Na^+. NBCs are crucial to regulate HCO_3^- absorption/secretion, HCO_3^- metabolism, pH_i regulation and regulation of cell volume. The electrogenic NBC1 catalyzes HCO_3^- fluxes in mammalian kidney, pancreas and heart cells [ALVAREZ *et al.*, 2003]. A characteristic acidic cluster of amino acids (DNDD) in the cytoplasmic C-terminal region of NBC mediates association with the basic N-terminal region of CAII, creating a functional complex or "transport metabolon" to facilitate HCO_3^- export from the cell [STERLING *et al.*, 2001, GROSS *et al.*, 2002].

At physiological pH_o, pH_i was in certain cell lines found to be higher in the presence of HCO_3^- than in its absence [BIERMAN *et al.*, 1988, CASSEL *et al.*, 1988, L'ALLEMAIN *et al.*, 1985]. L'ALLEMAIN (1985) showed that pH_i of Chinese Hamster Lung fibroblasts, in a bicarbonate-free medium, is regulated by both Na^+-coupled antiports (NHE and NCBE).

Cultured rat astrocytes were shown to have tree pH_i-regulating mechanisms in bicarbonate-containing medium: NHE, NCBE and AE [MELLERGARD *et al.*, 1993]. In his investigations, KRAPF *et al.* (1988) suggested that the cell pH defense against acute changes in pCO_2 depends on the basolateral NBC cotransporter (acid and alkaline loads) and the luminal NHE antiporter (acid loads). Additionally, the rate of the basolateral NBC cotransporter is a more important determinant of cell pH than the rate of the apical membrane mechanism.

CHAPTER 3

GENERATION OF A pCO_2 ROBUST CELL LINE

The generation of stable clonal cell lines ensures that the study of gene function can be conducted using a defined and homogeneous cell system. In this project it is intended to investigate the influence of the carbonic anhydrase II on cells' CO_2 metabolism of a cell line producing a recombinant protein of industrial interest. For this effect, genetic engineered mammalian cell lines stably expressing the human carbonic anhydrase II enzyme were generated. The metabolism of the cells was characterized and an ideal candidate was chose for this study.

3.1 STABLE TRANSFECTION OF HCAII GENE IN CHO CELLS

The human carbonic anhydrase II gene (hCAII) was cloned into the commercial available pIRES2-ZsGreen1 vector, generating the phCAII-ZsGreen1 plasmid (Section 10.1). A control restriction with BamHI was carried out in order to inspect existence and orientation of the hCAII-insert (data not shown). In the following subsections, the correct cloning of the hCAII is verified and, after transfection, the expression of the hCAII protein is detected by immunoblotting.

3.1.1 SEQUENCING OF HCAII GENE

The success of the cloning procedure was confirmed by sequencing the hCAII gene thus identifying the start and stop codon in the pIRES2-ZsGreen1 plasmid (Subsection 10.2.2). The result from base pair sequencing is presented in Figure 3.1.

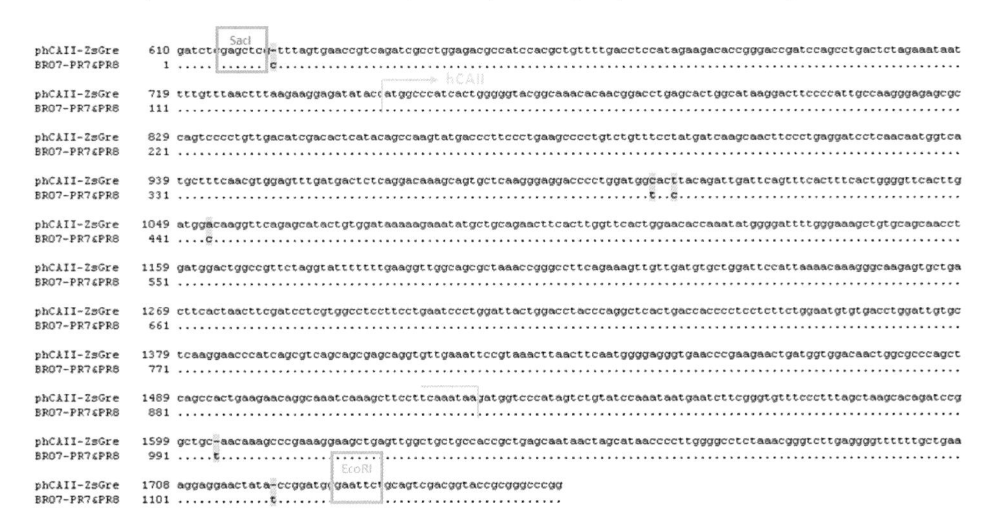

Figure 3.1: Sequencing of hCAII gene in pIRES2-ZsGreen1 vector. Comparison of the sequencing results (lower sequence: BR07) with the information provided by Dr. Casey (upper sequence: phCAII; Canada)

The hCAII gene is composed of 783 bp and the insert in phCAII-ZsGreen1 vector contains 126 bp 5´-untranslated region and 198 bp of 3´-untranslated region. The sequencing results (BR07) were compared by homology with the hCAII sequence provided by Dr. Casey (upper sequence; University of Alberta, Canada). It is possible to identify the restriction sites where hCAII was inserted in the vector backbone (SacI and EcoRI), and the start (atg) and stop codon (taa) of the hCAII gene. Three different base pairs are observed inside the hCAII gene sequence, at the positions 397 (c→t), 400 (t→c) and 445 (a→c) of the BR7 sequencing results. A global protein alignment was executed to verify if these punctual mutations have an influence in the amino acid sequence of hCAII (Fig. 3.2).

The reference protein sequence was the hCAII sequence from the Data Bank NCBI under the Reference Sequence NM000067 (upper sequence). The protein sequence length of hCAII is 260 amino acids. One change in amino acids was identified in

```
                                                                                         His64
                                                                                           ↓
hCAII_NM0000     1 mshhwgygkhngpehwhkdfpiakgerqspvdidthtakydpslkplsvsydqatslrilnnghafnvefddsqdkavlkggpldgtyrl
phCAII-ZsGre     1 .a........................................................................................
BR07-PR7&PR8     1 .a........................................................................................
                     His94↓ ↓His96                 ↓His119
hCAII_NM0000   271 iqfhfhwgsldgqgsehtvdkkkyaaelhlvhwntkygdfgkavqqpdglavlgiflkvgsakpglqkvvdvldsiktkgksadftnfdp
phCAII-ZsGre   271 ..........................................................................................
BR07-PR7&PR8   271 ..........................................................................................

hCAII_NM0000   541 rgllpesldywtypgslttppllecvtwivlkepisvsseqvlkfrklnfngegepeelmvdnwrpaqplknrqikasfk*
phCAII-ZsGre   541 ................................................................................*
BR07-PR7&PR8   541 ................................................................................*
```

Figure 3.2: Protein global alignment for hCAII gene. Upper sequence (hCAII): Data Bank NCBI (Ref. Nr.: NM000067; Middle sequence (phCAII): Dr. Casey sequence information (Canada); Lower sequence (BR07): Sequencing results). The active site of carbonic anhydrase formed by His64, His94, His96 and His119 is marked by arrows.

the sequence provided by Dr. Casey and in the sequencing results from the hCAII gene cloning. This mutation took place at amino acid position 2 and, therefore, it may not interfere with the catalytic center of the enzyme, which is composed of His64, His94, His96 and His119 (Fig. 3.2). It is not surprising that this mutation occurs in our sequencing results since the genetic material used for the hCAII cloning was provided by Dr. Casey. No additional amino acid mutations exist and the plasmid with the nucleic acid sequence encoding hCAII was further used for stable transfection.

3.1.2 TRANSFECTION AND SELECTION

For the investigation of the influence of the hCAII on the cells CO_2 metabolism, a cell line was generated with stable expression of the hCAII gene. Likewise importantly, a suitable control cell line was chosen, the non-transfected original cell line. Also to be considered is the DNA preparation for the transfection. Inside cells nucleus, plasmid DNA opens randomly and this can break some important plasmid features such as, the promoter or the polyA signal from the construct or even cause the disruption of hCAII gene during integration. And so, to have control where the plasmid is opened, the phCAII-ZsGreen1 plasmid was linearized with the restriction enzyme ApaLI (Subsection 10.1.3). This step may result in a decrease in transfection efficiency, but this should be made up for by an increased integration efficiency. For nucleofection, a CHO cell line culture was divided in three, one for transfection, the second as negative control and the third as positive control. The culture was nucleofected, as described in Subsection 10.3, with the linearized phCAII-ZsGreen1 plasmid, which also provides Geneticin resistance. Transgene expression was evaluated by FACS analysis (Subsection 10.9.1) 24 h post-nucleofection. The results are presented in Figure 3.3.

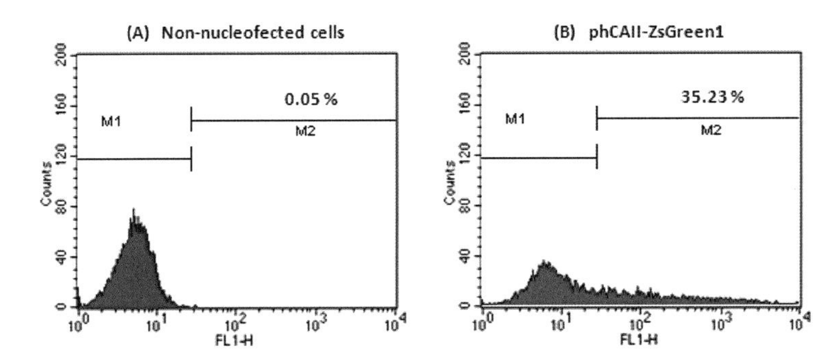

Figure 3.3: Nucleofection efficiency of U-24 program on CHO cells. CHO cells were nucleofected with the ApaLI-linearized phCAII-ZsGreen1 plasmid using the U-23 program and expression of the ZsGreen1 protein was detected 1 day post nucleofection by flow cytometry analysis for determination of transfection efficiency. (A) Control nucleofection (B) Nucleofected cells with the ApaLI-linearized phCAII-ZsGreen1 vector.

Before flow cytometric analysis, the control cell line had a viability of 88.1 % against 77.6 % for the nucleofected cells, indicating that the nucleofection with linear DNA induced higher mortality. FACS analysis of CHO cells 24 h after nucleofection showed 35.23 % ZsGreen1 positive cells. Nucleofection with circular DNA, conferred a nucleofection efficiency of 43.3 % (data not shown), supporting the assumption that nucleofection with circular DNA assures higher transfection efficiencies.

Clone selection and screening were done as explained in Subsection 10.6.2. Since the transfected cells contained a Geneticin resistance, selection was performed in HyQ medium with 400 $\frac{\mu g}{mL}$ G418 antibiotic. About 15 different single and double clones were isolated by limited dilution and further cultivated until reaching sufficient cells counts for western blot analysis and for freezing. The non-transfected cells served as negative control and were cultivated in selective medium until no more growth was observed. This negative control indicated when the selection procedure was over. In parallel, the control cell line was cultivated in HyQ medium, which did not contain methotrexate, during all clone screening experiment. On this way, differences were avoided due to different medium conditions and a comparable control was created. This positive control was designated as Control HyQ and the original cell line as Control. The later was always cultured in pre-culture medium containing methotrexate in its composition.

3.1.3 VERIFICATION OF HCAII EXPRESSION

After the transfection and selection, it was examined whether the expression of the hCAII gene was successful and substantially higher then carbonic anhydrase II expression of non transfected CHO cells. For this effect, western blottling was carried out with the polyclonal anti-CAII antibody as described in Subsection 10.10.5. In order to exclude some clones as early as possible, samples for immunoblot were taken during clone screening after selection was over. For this analysis, eleven clones were randomly chosen, among these, seven where single clones (Clones number 2, 3, 4, 7, 9, 10 an 11) and four clones came from two colonies (Clones number 1, 5, 6 and 8) (Table 3.1). Protein extraction of non-transfected cells was prepared as negative control. As positive control, the commercially available hCAII from Sigma was used. In Figure 3.4 the results of this experiment are shown.

Table 3.1: Clones used for immunoblot analysis. Single clones are marked with a star (*)

Clone number	Designation	Clone number	Designation
Clone 1	P01-D10	Clone 7*	P01-G9
Clone 2*	P08-C4	Clone 8	P01-E4
Clone 3*	P03-E8	Clone 9*	P01-E8
Clone 4*	P05-C7	Clone 10*	P10-G6
Clone 5	P01-C6	Clone 11*	P05-E11
Clone 6	P01-D6		

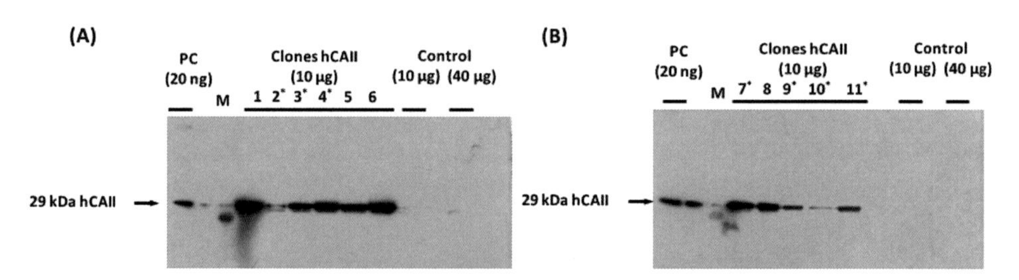

Figure 3.4: Immunoblot of hCAII in cell lines stably transfected with hCAII gene. (A) Immunoblot of hCAII clones 1 to 6 (10 μg). (B) Immunoblot of hCAII clones 7 to 11 (10 μg). Single clones are marked with a star (*). Control: non-transfected cells (10 μg or 40 μg); Positive Control (PC): carbonic anhydrase isozyme II from human erythrocytes (Sigma-Aldrich); Protein marker (M): SeeBlue® Plus2 Pre-stained standard (Invitrogen.)

For direct comparison of the results, the control cell lines proteome was applied with the same total protein amount as for the clones (10 μg) and with a 4-fold higher amount (40 μg). The different clones revealed different CAII expression levels as can be seen by the band thickness in Fig. 3.4. The detected 29 kDa CAII for the clones has the same size as the positive control carbonic anhydrase II. A small amount of protein dimer for the positive control (PC) was detected at 58 kDa with the immunoblot (Fig. 3.4 A) although reducing conditions were used. No CAII was detected for the non-transfected cells at different protein amounts applied. Relevant for this work are the stable cell lines generated from single colonies, as they provide a defined and homogeneous system. For that reason, the mixed clones will not be further considered. Five single clones were then classified as high-carbonic anhydrase II expressing cell lines (Clones 3, 4, 7, 9 and 11) and the other two as low CAII expressing cell lines (Clones 2 and 10). One of the CAII high producer clones, the Clone 9 (P01E8), died during the clone screening procedure. Therefore, from now on the clones designation is Clone E8, C7, E11, G9, C4 and G6 (Table 3.1). During cell propagation, ZsGreen1 expression drastically diminished as observed by fluorescence microscopy but hCAII-expression was conserved as confirmed by western blot analysis (data not shown). The results showed that hCAII was overexpressed in cells stably-transfected with phCAII-ZsGreen1 plasmid. In this experiment, as the same protein amount was applied in the SDS-Gel, a classification of the clones based on the hCAII expression was possible.

3.2 CHARACTERIZATION OF CLONES

For the examination, whether the expression of human carbonic anhydrase II had a significant influence on cells metabolism, parallel batch cultivations were performed (Subsection 10.6.1). Therefore, the different clones were compared with controls regarding some specific metabolic parameters, such as, growth, specific substrate rates and product formation.

3.2.1 GROWTH AND METABOLISM

The progression of the viable cell density and viability are shown in Fig. 3.5 *A* and *B*, respectively. The mean specific growth rates are illustrated in diagram *C*.

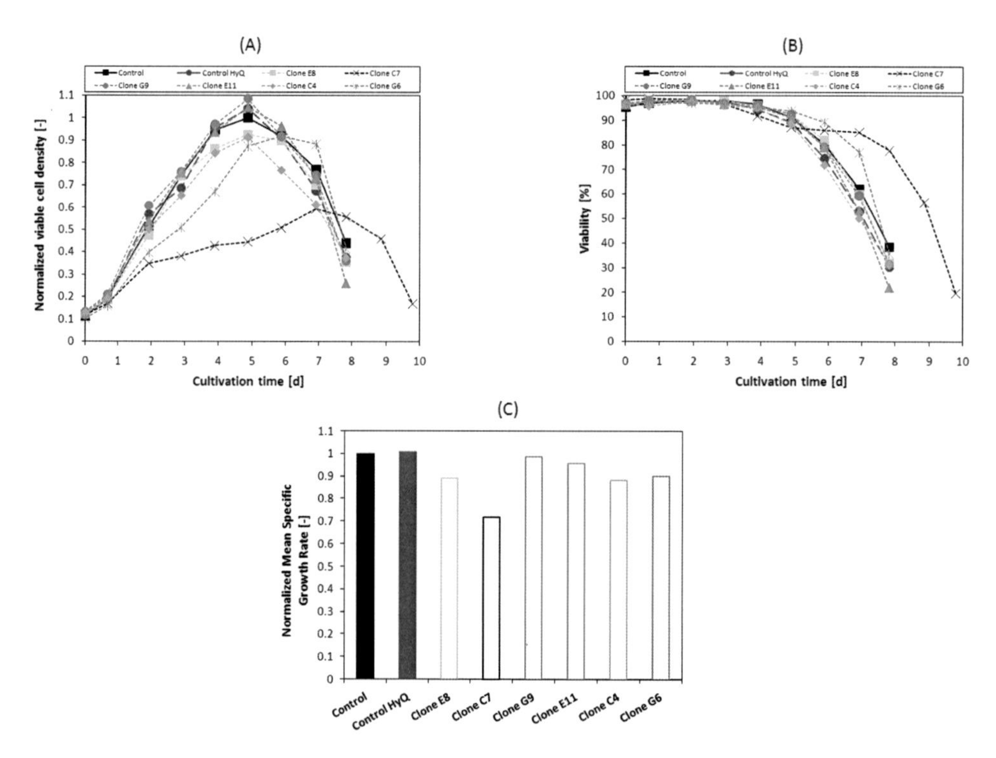

Figure 3.5: Batch cultures of Clones and Controls in 250 mL shaking flasks with 90 mL working volume in production medium. (A) Normalized viable cell density; (B) Viability; (C) Normalized mean specific growth rate.

In this experiment, two controls were included, *Control* and *Control HyQ*. Viable cell densities were normalized to the maximum viable cell density reached by *Control* on cultivation day 5. In the diagram *C*, the mean specific growth rates were normalized to *Control*. The maximum cell concentrations attained by the controls were comparable to the ones reached by Clones *G9* and *E11*. Clones *E8*, *C4* and *G6* achieved lower maximum cell densities then controls, although the last clone showed a lower evolution in growth and the maximum cell density was reached 1 day later. Clone *C7* grew very slow, possibly indicating that hCAII transfection had disrupted some important cell growth related genes. All batch cultivations, except for Clone *C7*, reached viabilities below 40 % after 8 days.

Except for the Clone *C7*, the growth in all examined cultures was comparable with growth in the range of $0.72\,\frac{1}{\text{day}}$ (Clone *G6*) and $0.84\,\frac{1}{\text{day}}$ (Clone *G9*). Only Clone *C7* showed a substantially lower specific growth rate of $0.57\,\frac{1}{\text{day}}$. *Control* and *Control HyQ* showed maximum growth rates of $0.79\,\frac{1}{\text{day}}$ and $0.88\,\frac{1}{\text{day}}$, respectively.

All clones except *C7* grew comparable to the original cell line, wether *Control* or *Control HyQ*. Development of the viability was also similar. Not taking into account Clone *C7*, the expression of the human carbonic anhydrase II by the CHO cells had neither a substantial influence on the maximum reached cell density in the batch cultures nor on the mean specific growth rate.

3.2.2 GLUCOSE AND LACTATE METABOLISM

The progression of the glucose and lactate concentrations in the culture medium during the batch cultivation are presented in Fig. 3.6 *A*. The starting glucose concentration was of about 18 mM for controls and clones, except for clone *C7*, and glucose depletion occurred at cultivation day 5. As already discussed in Subsection 3.2.1, the clone *C7* exhibit a slower growth and the result of the lower metabolism left more glucose available. In the cultivation of this clone, glucose concentrations above 0.5 mM were present until cultivation day 7. Until cultivation day 2, the lactate concentration rose in the clones cultures equally sharply as in the control cultures. The maximum lactate concentrations reached for the controls was 16 mM on cultivation day 3, while for the clones higher concentrations, of about 17.4 mM to 19.3 mM, were observed, between cultivation days 3 and 4. Afterwards, this undesirable metabolic product was consumed by the cells until the end of the batch culture. It is interesting to observe that when only lactate was available as a carbon source, the cells were unable to grow or maintain their viability. Lactate consumption was observed, which suggests that the cells cannot perform gluconeogenesis from pyruvate or obtain enough energy from lactate [ALTAMIRANO *et al.*, 2006].

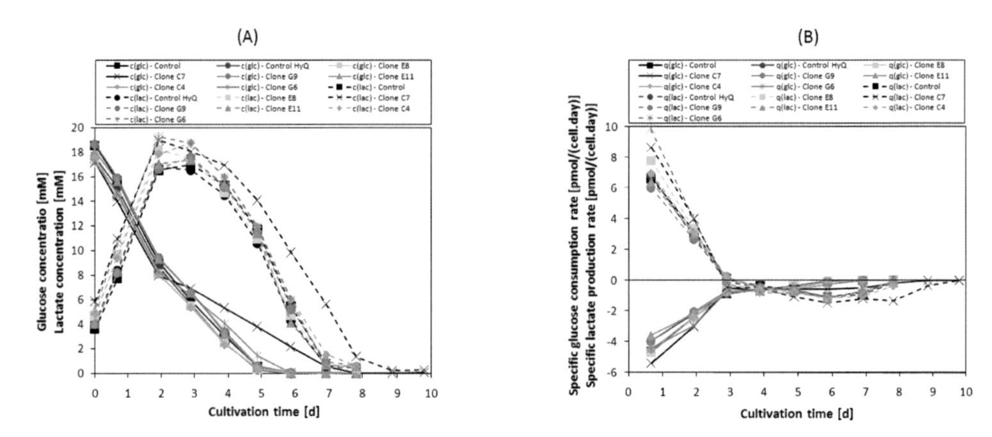

Figure 3.6: (A) Glucose and Lactate concentrations; (B) Cell specific Glucose consumption and Lactate production rates.

The calculation of the substrate consumption and metabolic product formation rates was made taking into account both, the exponential growth phase, where the consumption rates achieve their maximum values, and the course of cultivation, where the consumption rates decreased with the exhausting of the nutrients in medium [FITZPATRICK *et al.*, 1993]. Furthermore, the medium composition at the beginning of the cultivation in all cultures is quite uniform. In the course of the cultivation, differences in medium composition arise, leading to changes in the substrate consumption rates. Thus the consumption rates are not only dependent on the cell characteristics, but also on the culture conditions, which is not desired in the context of this experiment.

The cell specific initial consumption rates referred to pmol of substrate or metabolic product consumed or produced per cell and per day (Fig. 3.6 *B*). They were for glucose between 3.6 $\frac{\text{pmol}}{\text{cell.day}}$ (Clone *E11*) and 4.7 $\frac{\text{pmol}}{\text{cell.day}}$ (Clone *E8*) for the different hCAII clones. The values for the controls lie between those for clones and are 4.7 and 4.0 $\frac{\text{pmol}}{\text{cell.day}}$ for *Control* and *Control HyQ*, respectively. During the cultivation, the Clone *C7* had a slightly higher cell specific glucose consumption rate (5.4 $\frac{\text{pmol}}{\text{cell.day}}$) then the other clones and controls. This can be explained by the smaller initial substrate consumption leading to more glucose available in the culture medium, and therefore, to higher specific consumption rates. Initial cell specific initial production rates for lactate were between 6.0 $\frac{\text{pmol}}{\text{cell.day}}$ (Clone *G9*) and 7.8 $\frac{\text{pmol}}{\text{cell.day}}$ (Clone *E8*). Calculated for the controls were production rates of 4.7 and 4.0 $\frac{\text{pmol}}{\text{cell.day}}$ for *Control* and *Control HyQ*, respectively. Clones *C7* and *G6* exhibited substantially higher rates of 8.7 and 9.9 $\frac{\text{pmol}}{\text{cell.day}}$, respectively. A slower growth of the these clones together with initially observed higher specific glucose consumption rates, lead to higher initial lactate production and therefore higher specific consumption rates.

3.2.3 AMMONIUM METABOLISM

With the amino acids metabolism, ammonium ions accumulate in the culture medium. The evolution of ammonium ions accumulation in medium for control and clones and corresponding cell specific production rates are presented in Figures 3.7 *A* and *B*.

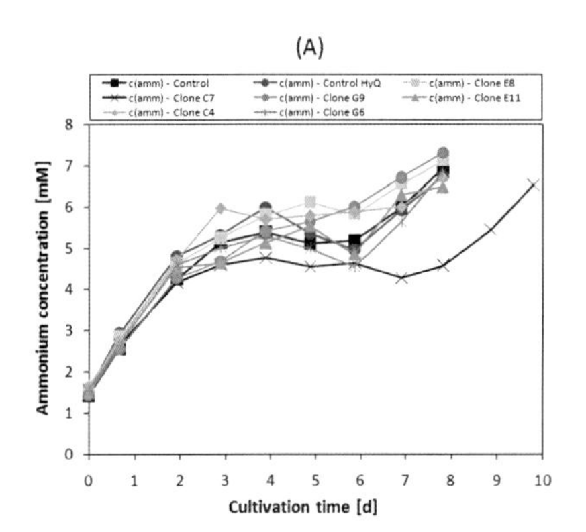

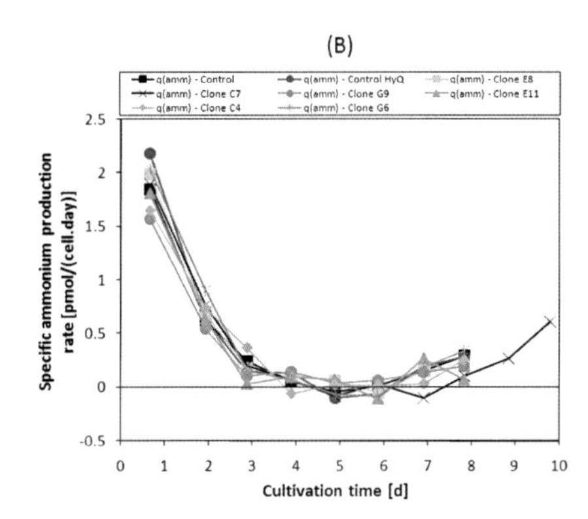

Figure 3.7: (A) Ammonium concentrations; (B) Cell specific Ammonium production rate.

The ammonium concentration rose both in the controls and clones cultures in a similar way, except for Clone *C7*, and reached finally concentrations between 6.5 mM (clone *E11*) and 7.3 mM (clone *G9*). For the clone *C7* a lower ammonium accumulation in medium was first observed after cultivation day 4, although a discrepancy in growth of this clone compared with the other clones and control was noticed already on day 2 of the batch culture. At the end of the batch culture, Clone *C7* reached an ammonium concentration of 6.5 mM, which is as high as for the other cultures.

The different clones initial cell specific ammonium production rates were between 1.57 $\frac{\text{pmol}}{\text{cell.day}}$ (Clone *G9*) and 2.03 $\frac{\text{pmol}}{\text{cell.day}}$ (Clone *G6*). The values for the *Control HyQ* is higher at 2.18 $\frac{\text{pmol}}{\text{cell.day}}$, while the *Control* ammonium specific production rate lay within the range of those obtained for the clones (1.84 $\frac{\text{pmol}}{\text{cell.day}}$). The progression of ammonium specific production rate was analogous for all cultures, rates decreasing to values almost nearly zero until cultivation day 4. After which a minor increase was observed at the end of the batch cultures. Among others, this effect may be due to the transamination of alanine at the end of cultivation.

3.2.4 PRODUCT FORMATION

Interestingly to observe is the influence of hCAII expression on product formation. The evolution of product formation and its cell specific productivity are represented in Fig. 3.8.

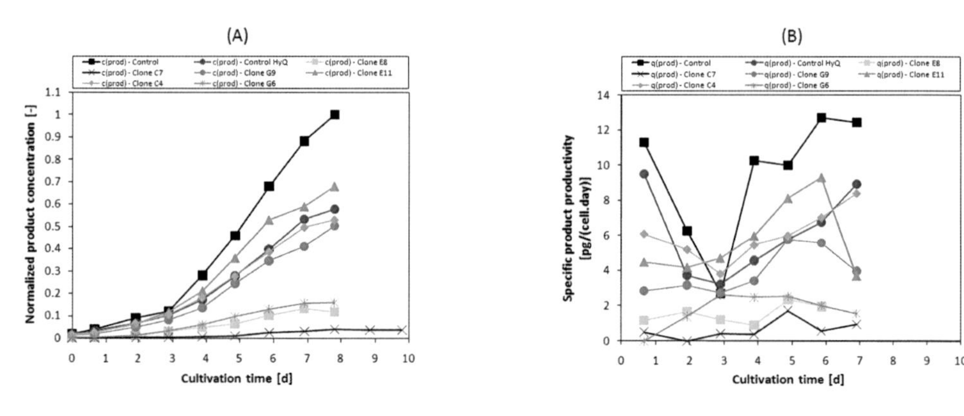

Figure 3.8: (A) Normalized product concentration; (B) Cell specific productivity.

After cultivation day 3, the product concentration increased linearly with time until day 7 in all cultures (Fig. 3.8 *A*). The maximum product concentration was reached for the *Control* at the end of the batch culture. In comparison to this result, only 58 % product was produced by the control cell line *Control HyQ*. Analogous product concentrations of 68 %, 53 % and 50 % of *Control*, were achieved by Clones *E11*, *C4* and *G9*, respectively. The other 3 clones achieved lower then 80 % of product concentration observed for *Control*.

For *Control*, the cell specific productivity increased after cultivation day 3 and reached its maximum after 6 days with 12.7 $\frac{\text{pg}}{\text{cell.day}}$ (Fig. 3.8 *B*). The *Control HyQ* and clones *E11*, *C4* and *G9* also had a similar product productivity progression, but associated to lower product productivities of 8.9 $\frac{pg}{cell.day}$ (day 7), 9.3 $\frac{pg}{cell.day}$ (day 6), 8.4 $\frac{pg}{cell.day}$ (day 7) and 5.8 $\frac{pg}{cell.day}$ (day 5), respectively.

The clone screening procedure had a significant negative influence on the cells' product formation. During the clone screening, the clones and *ControlHyQ* were maintained in HyQ medium for about 1.5 month. Methotrexate was not included in this medium formulation to facilitate clone screening and DHFR and product gene amplification using methotrexate was not active. With this results, the *ControlHyQ* will be used as control cell line for further experiments.

3.3 MYCOPLASMA STERILITY TESTING WITH PCR

The PCR amplification method following JUNG *et al.* (2003) was used for mycoplasma sterility testing in the supernatant from cultured cells. The results are presented in Fig. 3.9.

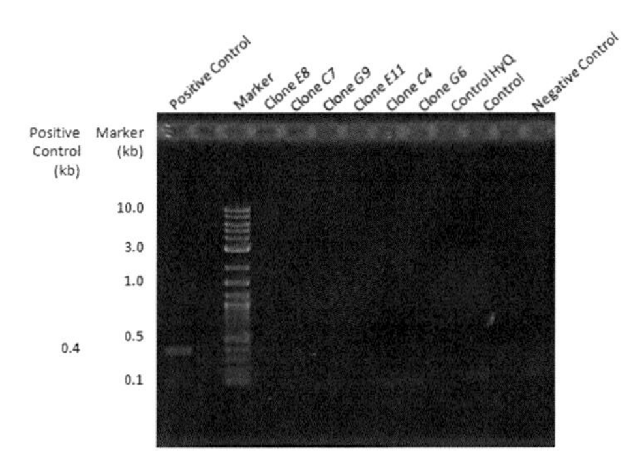

Figure 3.9: Mycoplasma testing by means of a PCR amplification method according to JUNG *et al.* (2003). Positive Control: mycoplasma infected cells supernatant; Marker: 2-Log DNA Ladder (0.1-10.0 kb; NEB)

Among the 8 cell lines analyzed, mycoplasma contamination was not detected using this method. Positive control exhibit an amplified fragment of about 400 bp and therefore, results indicate that none of all cell lines were infected, at least, with the same specie that infected the positive control.

3.4 hCAII Activity

From this first series of experiments, it is interesting to further investigate if the human carbonic anhydrase II is active in the clones that are metabolically similar with *Control HyQ*. Values of CAII activity were determined from the ^{18}O-exchange data obtained by mass spectrometry (Subsection 10.10.6). Through the titration curve of the CAII positive control solution with the tight inhibitor ethoxyzolamide (EZA), a CAII concentration of 1.33 μM was calculated. The determined activity of the non inhibited positive control was of 28700 units. Carbonic anhydrase II activity was only detected in the hCAII stably transfected cells lysate in RIPA buffer with activities of 201 and 241 units for clone *G9* and *E11*, respectively. The calculated CAII concentrations were of 2.28 nM (clone *G9*) and 2.73 nM (Clone *E11*). Lysates resultanting from protein extraction by mechanical disruption in TE buffer provided CAII activity only for clone *E11* with 1235 units, corresponding to a CAII concentration of 14.0 nM. No CAII activity was measured for clone *G9*. None of both extraction methods provided detectable CAII activity in *Control* or in *Control HyQ* using the present enzyme activity test. A possible explanation for the not detectable activity may be the very low CAII concentrations present in the samples. Enough carbonic anhydrase II was present in lysates from protein extraction of clone *E11* derived from two different protein extraction methods and enzyme activity was detected on both samples. These results, together with the metabolic characterization results, makes clone *E11* a very suitable hCAII stable transfected cell line to analyze the short and long term cytoplasm acidification on next experiments.

3.5 Interpretation of Results

A cell line was generated that stably expresses the human carbonic anhydrase II. The fact that a different amino acid in the sequence exists, apparently didn't had an effect on protein expression or activity since both tests were positive. These results suggest that the transfection via nucleofection of linear DNA raise lower transfection efficiencies. The different geometry of the two topologies might play a role in the disparity in the transfection efficiency. Comparisons between circular and linear DNA have been performed for nucleofection of CHO-K1 an K562 cell lines. A slight tendency for higher nucleofection efficiencies was obtained for the circular DNA form [SCHMIDT *et al.*, 2004]. Circular DNA tend to form compact aggregates and the linear plasmid, necklace-like structures, making more difficult to cross cell wall [VON GROLL *et al.*, 2006]. The previous DNA linearization aid the integrity of the human carbonic anhydrase gene and the prevention of selection of false clones. Some authors are even in the opinion that linear DNA nucleofection most probably lead to higher integration efficiencies due to the slightly higher rates of resistant cells observed [SCHMIDT *et al.*, 2004]. What could not be predictable was the random integration of the plasmid in cell's genome which might disrupt important genetic pathways. Gene transfer via transfection with linear DNA may be favored to obtain better transfection efficiencies but, the DNA geometry on the influence on the creation of cell line clones characterized by a single integration event may be different. In the present work, this latter aspect was not investigated. In some cell lines, a desired single integration event may be favored by nucleofection of circular rather than linearized plasmid [SCHMIDT *et al.*, 2004].

The sequencing results showed that the protein sequence is intact besides one base alteration at the beginning of the amino acid sequence (position 2). The initial and end codon are present and, therefore, it was expected that the expressed protein had the right size. This was confirmed by immunoblotting. Since active center sequence didn't suffer any alterations the protein should have the right conformation to be active. Activity tests demonstrated that the expressed protein is active.

The expression of the human carbonic anhydrase II was successfully detected by immunoblot analysis for all examined CHO-hCAII clones. The size of the detected protein was of 29 kDa, although other authors obtained a bigger carbonic anhydrase II (around 35 kDa) with the same hCAII sequence (pJRC36, the same vector) transfected in CHO cells [LI *et al.*, 2002]. The expected size for the human CAII is 29246 Da (SwissProt - hCAII Ref N°. P00918), which is in agreement with the results obtained. The different expression levels were distinguishable by the different bands thickness. This difference arise from the additional expression of the exogenous human CAII since no detection was achieved for the endogenous hamster carbonic anhydrase II. The antibody used for this immunoblot is known to react with human and mouse and not cross react with the carbonic anhydrase I (hCAII Cat. N.° ab6621, Abcam). Although, a detection of the endogenous carbonic anhydrase II from hamster was already done by others [LI *et al.*, 2002] and in this project Section 5.4. The CAII antibody is polyclonal and therefore recognize multiple epitopes. Carbonic anhydrases are highly conserved across mammals and a cross reaction with the hamster carbonic anhydrase II might be possible. The carbonic anhydrase present in the marker was also detected with this antibody. In the conditions of the electrophorese gel performed in this work, the CAII present in the marker should present a size of 28 kDa. No information could be obtained about the origin or type of the carbonic anhydrase present in the marker.

During the clone screening procedure, an undesired loss of the ZsGreen1 expression occurred. The idea was to follow the human carbonic anhydrase II expression through the ZsGreen1 fluorescence since the genes coding for both proteins are connected by a IRES element and are co-translated from a single bicistronic mRNA. The weak IRES-mediated expression of ZsGreen1 may be mostly attributed to the weakness of the IRES since the expression of the carbonic anhydrase II was detected by immunoblot. That effect was also detected by others [KUMAR *et al.*, 2003]. Due to the problems encountered during clones screening, a suitable parameter for the selection of the clones was hCAII expression detected by immunoblot. The single clones that expressed hCAII were further characterized in batch cultivations. Apparently, the expression of the hCAII had little or no effect on cells growth or metabolism. The insignificant differences can be attributed to the fact that it is a single experiment. In the other hand, recombinant protein production was considerable different for clones and controls. This negative impact was attributed to the conditions of the clone screening procedure. The long-term cultivation in medium without methotrexate may had be the cause for the drastic decrease on productivity. This effect was clearly seen in the productivity between controls. Due to this fact, the control cell line chosen was the original cell line cultured in medium without methotrexate. Some scientists use a clone mixture as control cell line, others decide on the transfection with an empty vector. In both ways, the integration of the plasmid occurs randomly and therefore it might happens that other genes are disrupted. In this way, other interferences in growth and/or metabolism may be unwantedly include in the control cell line. In this project it was important to exclude maximum divergences between control cell line and cloned cell line as possible. A hCAII-expressing clone will be further tested in bioreactor cultivations under different CO_2 profiles.

In conditions were galactose is present in the culture medium, like in these experiments, the by the cells produced lactate was re-metabolized. Some authors also reported partial lactate assimilation in other animal cell cultures [OZTURK *et al.*, 1997] and other the total assimilation in CHO cells cultures [ALTAMIRANO *et al.*, 2004] in the presence of galactose.

During all process of generation of the hCAII expressing clonal cell lines, starting from nucleofection, going through limited dilution and ending with clone screening, the risk of contamination is very high. The process is very time consuming, it involves the addition of some reagents and the cultivation of the cells in diverse cultivation systems. Mycoplasma contamination is a silent and neglected danger. The detrimental effects of mycoplasma contamination on cell lines has been described in detail and are a major problem in cell culture [LINCOLN AND GABRIDGE *et al.*, 1998; KAGEMANN *et al.*, 2005; MIRJALILI *et al.*, 2005]. The clonal cell lines and the controls were authenticated in terms of mycoplasmal contamination. Culture medium was used as the source of PCR templates as tested by JUNG *et al.* (2003). Using this method, no contaminations were detected in any of the 8 clones neither in the control cell lines. Taking into account this and the metabolism results, one can ensure valid and reproducible experimental results in further experiments with those cell lines.

Measurements with the ^{18}O-exchange method showed that the expressed hCAII is active in two clones. This conclusion was further supported for one of the clones, Clone E11, by the detection of activity in other protein extraction solution. For the control cell lines and for the other clones, no activity was detected. This result may be explainable by the low hCAII enzyme concentration in those sample. A concentration of carbonic anhydrase II on the samples may be useful to obtain higher concentrations that are detectable by this method. The main advantages of the ^{18}O-method are that it is a sensitive method for the detection of the carbonic anhydrase activity [VINCENT AND SILVERMAN, 1982]. Most of the scientists don't verify is the transfected hCAII is active, instead of that they use indirect measurements of transporters activity of the co-transfection of those transporters with the carbonic anhydrase through pH_i change. Some examples are the co-transfections of the CAII with the NHE1 [LI *et al.*, 2002; 2006], with AE1 [STERLING *et al.*, 2001] and with NBC1 [ALVAREZ *et al.*, 2003; LU *et al.*, 2006]. These authors only determined the hCAII expression by immunoblot. The amino acid mutation seems not to have any influence on hCAII activity.

First it should be mentioned that the transfection procedure and the subcloning play a large role for the selection of the correct clone that expresses the desired protein. In order to chose a clone that is metabolically equal/similar with the original cell line, several hundred clones would have to be screened and many other involved parameters should be also be taken into account. This is connected to months work. In the context of this project only 12 random clones were selected and more near analyzed, therefore it can be assumed that the expression of carbonic anhydrase correspond to an average value. The only satisfactory clone demonstrating the same growth and metabolic characteristics as the original cell line Control HyQ and expressing an active hCAII was the Clone E11. This hCAII-expressing cell line and the adequate Control HyQ will be in the next part of this project further characterized.

CHAPTER 4

PHYSIOLOGIC EFFECT OF CAII ON pH_I

A possible mechanism of inhibition by elevated carbon dioxide is through disruption of the intracellular pH value. The carbonic anhydrase II enzyme is directly involved in the regulation of this intracellular parameter. Therefore, the stable cell line generated in the first part of this project was tested for the physiologic effect of the expression of the human carbonic anhydrase II gene on the pH_i under CO_2 acid load.

4.1 ACID LOAD

To examine the effect of hCAII overexpression on the activity of the pH_i, *Control HyQ* and clone *E11* were submitted to a short term acidification. This was achieved by mixing the cells with a CO_2-containing solution and measuring the subsequent pH_i recovery as described in Subsection 10.9.3. Fig. 4.1 *A* illustrates the evolution of pH_i from an acid load in presence (clone *E11*) or absence (*Control HyQ*) of hCAII and Fig. 4.1 *C* shows the corresponding CO_2 concentration during the experiment. A replicate of this experiment was performed with the carbonic anhydrase II inhibitor acetazolamide and the results are presented in Figs. 4.1 *B* and *D*. The initial rate of recovery and the results normalized for *Control HyQ* are demonstrated in Figs. 4.1 *E* and *F*. Table 4.1 summarizes some of the characteristics of the tested cell lines.

The response of the intracellular pH to cytoplasm induced CO_2 acidification was consistent for all experiments with both cell lines. A stronger decrease of the pH_i was

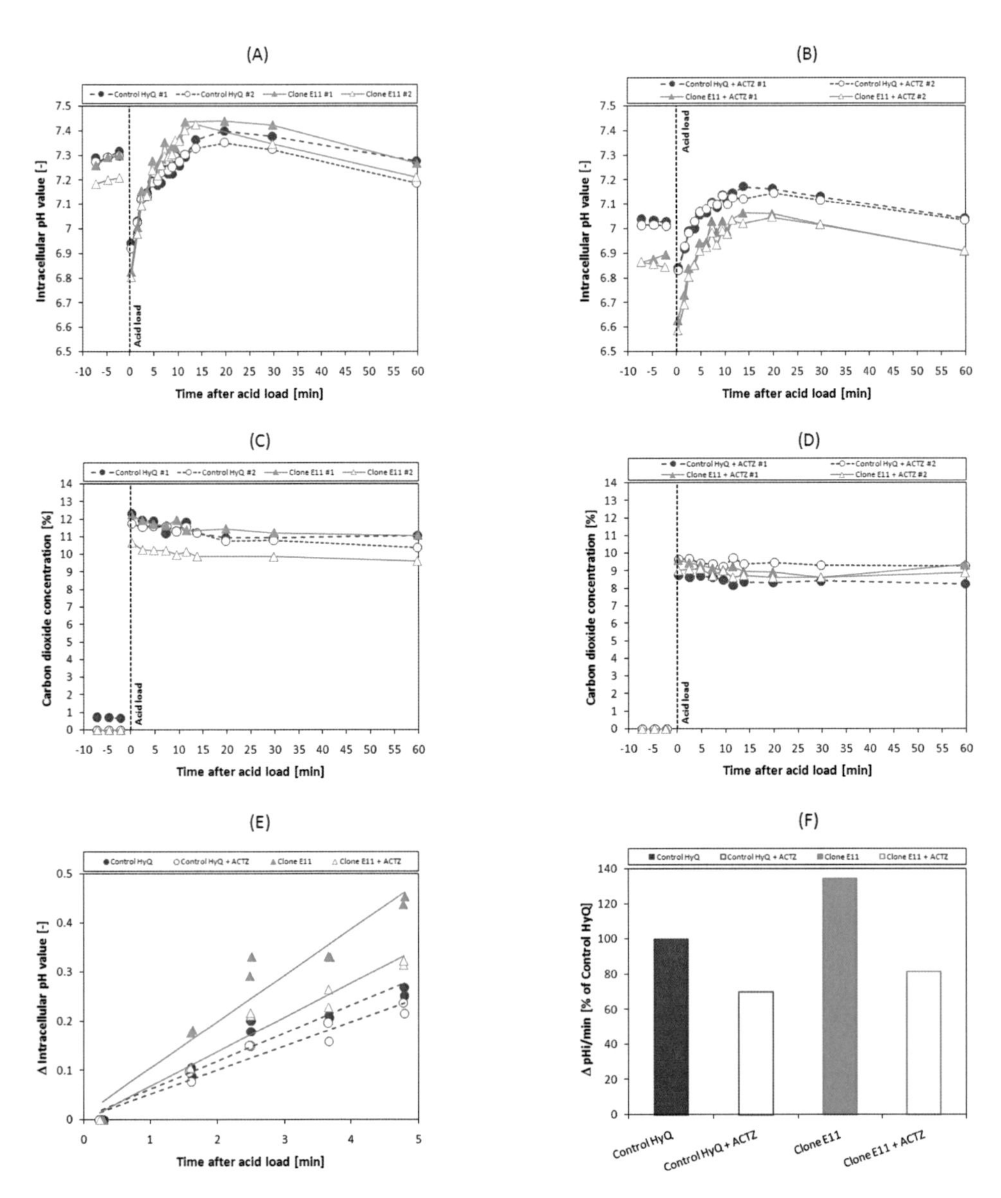

Figure 4.1: Effects of CAII expression and acetazolamide on pH_i of the Clone E11 and control cell line after acid-load. (A) Intracellular pH recovery after acid-load; (B) Intracellular pH recovery after acid load in the presence of 100 μM acetazolamide; (C) Carbon dioxide concentration during the acid load experiment; (D) Carbon dioxide concentration during the acid load experiment in the presence of 100 μM acetazolamide; (E) Initial pH_i recovery after acid-load; (F) Comparison of the initial pH_i recovery rates.

Table 4.1: Characterization of the original cell line (*Control HyQ*) and stable cell line expressing hCAII (clone *E11*). Steady state pH is the pH before the acid load. Results are mean values from two experiments.

	Control HyQ	**Clone *E11***	**Control HyQ + ACTZ**	**Clone *E11* + ACTZ**
pH_e	7.18	7.18	7.10	7.11
Steady state pHi	7.29	7.24	7.02	6.87
pH_i (acid load)	6.93	6.81	6.84	6.61
$\frac{pH_i}{min}$ (5 min)	0.070	0.094	0.049	0.057
r^2	0.934	0.938	0.947	0.959

observed for the hCAII-expressing cell line compared to the control (with or without inhibitor). Afterwards, the recovery started, indicating regulation of the pH_i. The clone *E11* evidenced quicker pH_i changes which were linked to greater pH_i variations than observed for the *Control HyQ*. An additional increase of pH_i above the steady state value was observed 8 min after acid load for clone *E11* as opposed to 11 min for control cell line without inhibitor. For the CAII inhibited reactions, the steady state was exceeded 5 min after exposure of cells to CO_2-containing medium for both cell lines. Sixty minutes after cytoplasm acidification, the steady state was restored.

Carbon dioxide (Figs. 4.1 *C* and *D*) and extracellular pH (data not shown) were maintained rather constant during all experiment due to the closed system used. In the experiment without inhibitor, the transient acid load brought the cells in contact with ca. 12 % CO_2, except for test number 2 for clone E11 (11 % CO_2). For the CAII inhibited experiments, this value was around 9 % CO_2. A difference of 0.1 pH unit was observed for the pH_e between tests with or without CAII inhibitor. The steady state pH_i before the acid load in both cell lines was different in all experiments for the diverse conditions (with or without acetazolamide), being the lowest 6.87 pH units for the trial with clone *E11* with acetazolamide. Despite this, the external pH was the same as for the control.

The recovery rates of the cells were calculated for the linear response of the first 5 min after acid load. In this period of time, the clone *E11* revealed a 34 % increased initial recovery rate, compared to the *Control HyQ*. This is indicative of stimulation of pH_i regulation mechanisms by CAII overexpression. This tendency was still maintained when the CAII inhibitor was present but associated to lower initial recovery rates indicating that the CAII inhibition by acetazolamide (100 μM) limits substrate availability the

pH_i regulation mechanisms. A 30 % and 19 % decrease on the re-alkalinization rates were calculated for the *Control HyQ* and clone *E11*, respectively, in the presence of the inhibitor.

4.2 DISCUSSION OF THE RESULTS

Direct variation of pH_i was used to analyze the effect of CAII overexpression on pH_i regulation. The pH-sensitive fluorochrome SNARF-1 AM have been used to evaluate pH_i by measuring the fluorescence emitted by single cells with ratiometric methods, such as flow cytometry [LI *et al.* 2002, LI *et al.*, 2006 (acidification by CO_2), DOLZ *et al.*, 2004 (acidification by propionic acid in bicarbonate-free medium)].

In this part of the work the pH_i recovery after acid load was examined when the human carbonic anhydrase II was overexpressed. This is the first work to report the re-alkalinization rates of hCAII-transfected cells. It was demonstrated the feasibility of using flow cytometry and off-line gas analysis to monitor the steady-state pH_i as well as rapid pH_i changes in mammalian cells. Using this technique it was shown that hCAII overexpression improved the initial recovery rate after carbon dioxide acid-load. The hCAII transfected cell line exhibited the highest alkalinization rate against the control cell line in the same conditions. Addition of hCAII-inhibitor decreased the initial rate of recovery by greater than 17 % in both control and clone cell lines. These effects were already reported by others [LI *et al.*, 2002] but the experience was done with NHE1-hCAII-transfected cell line. Since ABSTON AND MILLER (2005) reported that NHE1 overexpression doesn't bring any benefit to cells growth at elevated pCO_2 levels (195 mmHg pCO_2/435 $\frac{mOsmol}{kg}$) then most probably only CAII was stronger involved in the re-alkalinization mechanism. Therefore, our results are consistent, ie., in the same range of values, with the ones obtained by LI *et al.* (2002). Differences on the absolute pH_i values may be attributed to differences of the pH_e and/or calibration curve that was not done specifically for this experiment but the same cell line and settings were used. Because pH_i is influenced by pH values in intracellular compartments, no absolute measures of pH_i can be made without perturbing the cell. Therefore, there is greater confidence in measuring changes in pH_i than in determining an absolute value [CHOW AND HEDLEY, 1997].

The faster answer of hCAII transfected cells to an acid load lead to a higher increase of pH_i (above 7.4) than for non-transfected cells and after 60 min the pH_i was equal for both cell lines. LI *et al.* (2002) only reported the first minutes of the re-alkalinization experiment, not allowing a comparison of the results. JOCKWER (2008) also observed the same type of effect but for re-acidification of cells. Indeed, after 40 min the pH_i was still decreasing below the initial pH_o.

LI *et al.*, (2006) reported that steady state pH_i of cells with or without CAII was not altered and suggested that the physiological role of CAII is thus more related to the recovery from exposure of cells to weak acid in the form of CO_2, rather than steady state pH_i maintenance. It is of note that, with acidification of cells by ammonium chloride in the absence of HCO_3^-, there was no difference in the rate of recovery between the cells with or without CAII. This supports the suggestion that carbonic anhydrase activity is necessary for the facilitation of NHE1 exchanger activity with exposure of cells to weak acid in the form of CO_2 and that carbonic anhydrase does not simply stimulate NHE1 activity in a nonspecific manner.

The basal and regulated transport rate is dependent on the number of active exchangers expressed on the cell surface and on the turnover number of the individual exchangers [CAVET *et al.*, 1999]. The turnover number of a transport protein can be influenced by intrinsic properties of the protein itself and extrinsic properties such as cytoskeletal and signaling molecule interactions and membrane lipid fluidity [CAVET *et al.*, 1999]. The turnover rate for CAII ($\sim 10^6\,s^{-1}$) is very fast compared with most other enzymes [VOET AND VOET, 1995], this is unlikely to be rate-limiting step of the reaction. In contrast, the turnover rate for NHE1 is at least 3 orders of magnitude slower [VERKMAN AND ALPERN *et al.*, 1987; CAVET *et al.*, 1999], suggesting that NHE1 activity limits the export rate of H^+ ions by the CAII.NHE1 transport metabolon, thereby limiting the pH re-alkalinization of the cytoplasm.

In their re-alkalinization experiments, LI *et al.* (2002) only considered the NHE1 and ignored other transporters, including bicarbonate transporters, which are also crucial since the experiments were done in presence of CO_2. In Vero cells, NHE1 does not play a major role in maintaining pH_i at physiological pH_o since it displays a very low basal activity and it is mainly responsible for acid efflux after an acute severe acid load ($pH_i < 6.5$), whereas the Na^+-dependent HCO_3^--Cl^- antiport (NCBE) is responsible under milder physiological changes (pHi > 6.5) [TONNESSEN *et al.*, 1990]. The fact that NHE1 action alkalinizes the cell, and that NHE1 auto-inhibits at alkaline pH, is underappreciated [COUNILLON AND POUYSSEGUR, 2000]. Because in the present experiments the pH_i never dropped below 6.5 in our experiments, it is most likely that the NCBE played a more important role in the pH_i recovery. TONNESSEN *et al.* (1990) investigated deeply in this area and observed the activity of the NCBE over pH_i values tested from 6.4 - 7.4, with this antiport been more active at pH_i 7.0 than 7.2, therefore it will counteract acidification over a wide pH_i range. It can therefore be considered as a constitutively active antiport. On the other hand, the acidifying activity of the Na^+-independent HCO_3^--Cl^- antiport (AE) is strictly regulated by the pH_i and is activated when pH_i increases above 7.1 in Vero cells [TONNESSEN *et al.* (1990)]. This step activation of the AE antiport allows a fine tuning of pH_i. The curves obtained in Figs 4.1 are supported by this findings and may result therefore of the combined action of NCBE and the AE. The former antiport can transport H^+ out and HCO_3^- in, elevating therefore the pH_i (alkalinization) and, during the time pH_i raises, it became gradually less active, and the AE antiport gradually got more active. The NHE1 transport may then not be involved at all in the pH_i alkalinization at physiological pH_i.

The plasma membrane exchanger of erythrocytes, AE1, and hCAII interact with each other and forms a transport metabolon, a membrane protein complex involved in regulation of bicarbonate metabolism and transport. The binding of CAII to the AE carboxyl-terminus potentiates anion transport activity and allows for maximal transport [STERLING *et al.* 2001]. AE1 and CAIV interact on extracellular loop 4 of AE1, forming the extracellular component of a bicarbonate transport metabolon, which accelerates the rate of AE-mediated bicarbonate transport.

The overexpression of CAII may also cause an increase of freely diffusing CAII that in turn increases the rates of H+ and HCO3- diffusion by generating mobile bicarbonate buffers [SPITZER *et al.*, 2002], which tend to homogenize cytoplasm pH. Intracellular pH regulation is very complex and it involves many proteins and exchangers. CAII plays a very important role and interacts with some of those exchangers. How and what happens is still not totally clear but with these experiments, an overexpression of hCAII gives insights to enhance the answer to pH_i changes after an acid load with CO_2, at least in a short time it brings advantages to cells.

In the framework of this project, it is very interesting to look at the initial recovery rates. There are some limited data for mixing times at very low power inputs for large scale. Some scientists reported mixing times of typically 100 s (ranging from just greater than 200 s to approx. 70 s) for a 8000 L bioreactor for cell culture medium (NIENOW *et al.*, 1996). In such a bioreactor, the cells may experiment different CO_2 conditions due to hydrostatic pressure. So, the higher initial re-alkalinization rate from clone E11 may bring some advantages in such situations where mixing times are around 1-2 min, although the pH_i variation is higher.

CHAPTER 5

EFFECT OF LONG-TERM CO_2 INCREASE ON CELLS

In large scale cultivations, many parameters play an important role in culture performance that do not exist on bench top scale. The carbon dioxide distribution is one of these parameters and its accumulation to non-physiological levels might have a negative impact on cells. In this part of the project, the CO_2 accumulation in a large scale cultivation was simulated in bench top scale based on literature data [MOSTAFA AND GU, 2003]. The behavior of the human carbonic anhydrase II expressing cell line was investigated under these conditions.

5.1 pCO_2-CONTROLLED CULTIVATIONS

Parallel batch cultivations were performed in a controlled system as described on Subsection 10.6.3. It was tested if the hCAII-expressing cell line contributes with any advantage to cultivations at long-term increased CO_2 levels as opposed to cultivations controlled at physiological pCO_2. For comparison, the control cell line was also cultivated under the same conditions.

5.1.1 GROWTH

Until day 2, both cell lines were cultivated at a constant level of 5 % CO_2, after which a profile was imposed with gradual increase of CO_2 from 5 % to 25 % within 4 days. The reference culture was maintained at controlled 5 % CO_2. The cultivation data is illustrated in Fig. 5.1.

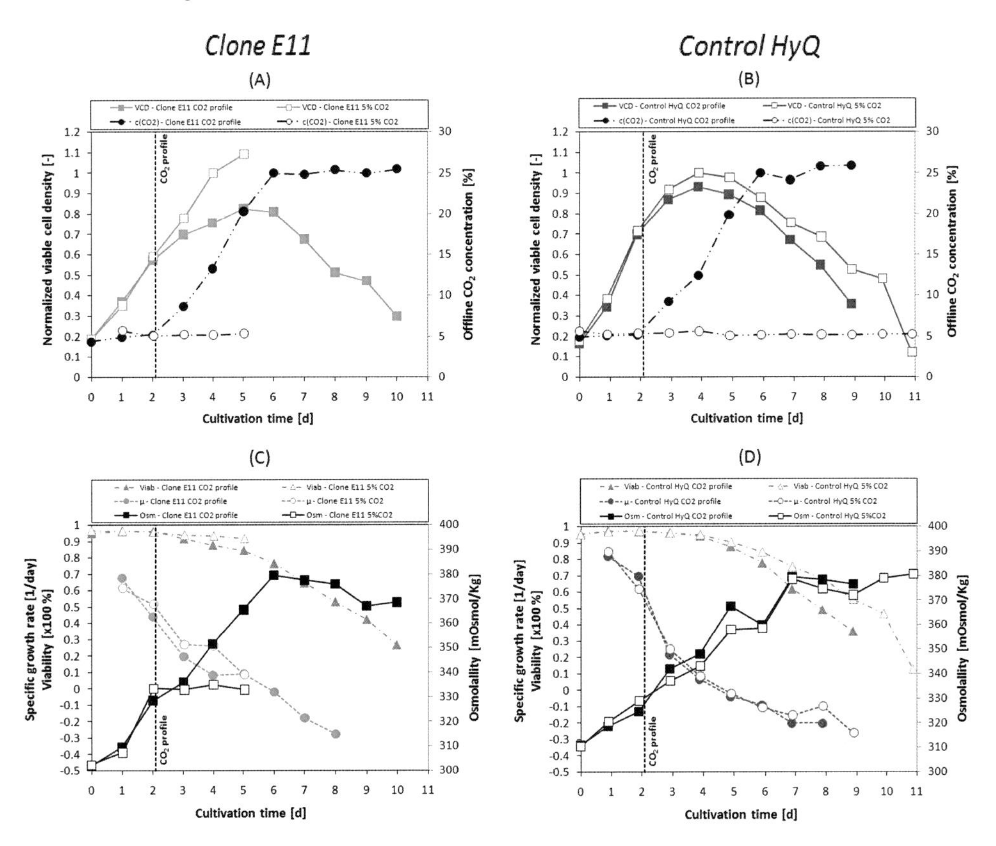

Figure 5.1: Batch cultivations of the hCAII-expressing cell line (right) and of the Control HyQ (left) in bioreactor with 1.5 L working volume in production medium. (A) and (B) Viable cell density and CO_2 profiles. (C) and (D) Viability, specific growth rate and osmolality.

The viable cell densities were normalized to the maximum cell density reached by the control cell line at cultivation day 4. Although the culture conditions were the same until day 2, a slower growth was observed for the Clone E11 compared with the original cell line. This growth was characterized by initial lower specific growth rates and before the CO_2 profile started, a 22 % increased viable cell density was reached for the Control HyQ compared to the clone E11. This behavior was not observed in the non-controlled shaker experiments (Section 3.2). After day 3, while the control cell line nearly reached a plateau, the hCAII-expressing cell line proceed growth at 5 % CO_2. At day 5, the clone E11 presented an 11 % increased maximum viable cell density. Unfortunately, at day 6 a contamination was identified for this culture and it was harvested. The result of the imposition of a CO_2 profile was a slower growth for both cell lines. The effect of the raised CO_2 level on the growth was considerably higher for the clone E11 than for the Control HyQ. With only one prolonged cultivation day, the maximum attained cell density was 12 % lower for the clone E11 when compared to the control cell line. A 24 % lower maximum viable cell density was reached for the cultivation of the clone E11 at different CO_2 profiles. Among other factors, differences in osmolality may have played a role in the better growth of the clone *E11* at 5 % CO_2. Although the osmolality was corrected with a NaCl solution after day 4, a non-efficient osmolality correction led to a discrepancy of 16 $\frac{\text{mOsmol}}{\text{Kg}}$. This effect doubled the following day. At the end of the batch of the control cell line, 380 $\frac{\text{mOsmol}}{\text{Kg}}$ was measured for the 5 % CO_2 controlled cultivation and 375 $\frac{\text{mOsmol}}{\text{Kg}}$ for the CO_2 profile cultivation.

5.1.2 PRODUCT FORMATION

The long-term controlled carbon dioxide increase had a negative effect on cells growth. The influence of this parameter on the cells recombinant protein production is illustrated in Fig. 5.2. The product concentration was normalized to the final product concentration reached by the Control HyQ in the physiological controlled CO_2 cultivation at the end of the batch. The clone E11 showed higher product concentrations during the cultivation process. A very interesting result is the increase of cell specific productivity for both cell lines at increased CO_2 levels. For the control cell line this divergence is observed after day 5 when the cultivation enters on the dying phase. While for the hCAII-expressing cell line this effect took place earlier on day 4 before the cells reached its maximum cell density. Regarding these results it is expedient to investigate the cells' metabolism, intracellular pH_i and cell cycle distribution.

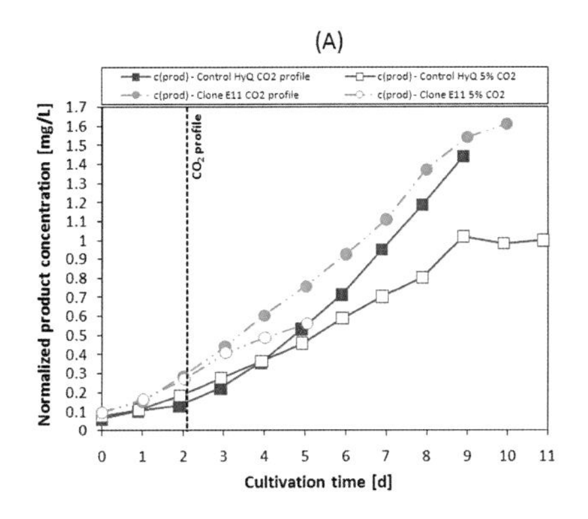

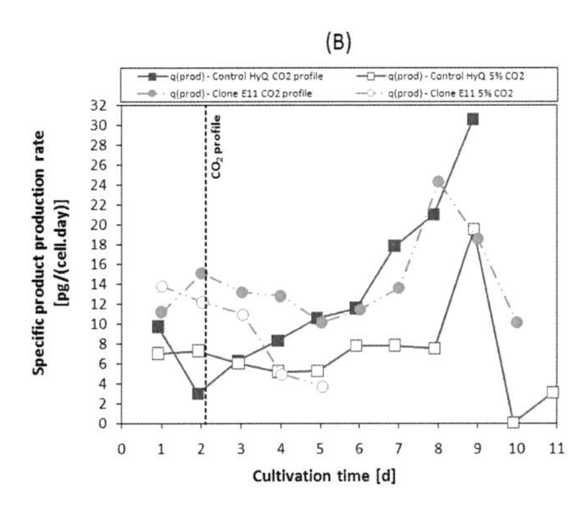

Figure 5.2: (A) Normalized product formation; (B) Product cell specific production rate.

5.1.3 METABOLISM

The influence of the different CO_2 profiles on the glucose consumption and lactate production are illustrated on Fig. 5.3. Differences to be noted are the lower lactate concentrations reached for the cultivation of the Control HyQ submitted to increased CO_2 levels. While for the hCAII-expressing cell line, 27.5 and 26.3 mM of lactate were measured at day 3 for the 5 % and increased CO_2 cultivation, respectively. After which, lactate concentration decreased until the end of the cultivation. No significant differences were observed on the cell specific consumption or production rates of glucose and lactate. Ammonium concentration and specific production rate evolution were similar for all cultivations. At the end of the batch culture, ammonium concentrations measured were of about 6 mM.

Only the amino acids that showed limitations during the batch culture were investigated (Fig. 5.4). All other amino acids were present in enough concentration in medium. In all cultures, a high consumption was observed for glutamine (GLN), asparagine (ASN) and tryptophan (TRP). Glutamine was the first amino acid to be limiting already on day 3 for all cultures. The non-essential amino acid asparagine and the essential amino acid tryptophan reached exhaustion on day 6. In the cultivation of the clone E11 under increased CO_2, lysine (LYS; essential amino acid) was present at concentration below 100 μM after day 8. On the other hand, aspartate (ASP), alanine (ALA), glutamic acid (GLU) and glycine (GLY) were produced (data not shown). Additionally, aspartate and alanine were re-metabolized at a certain point of the cultivation. Aspartate consumption was associated to lower levels of asparagine in the medium. No considerable

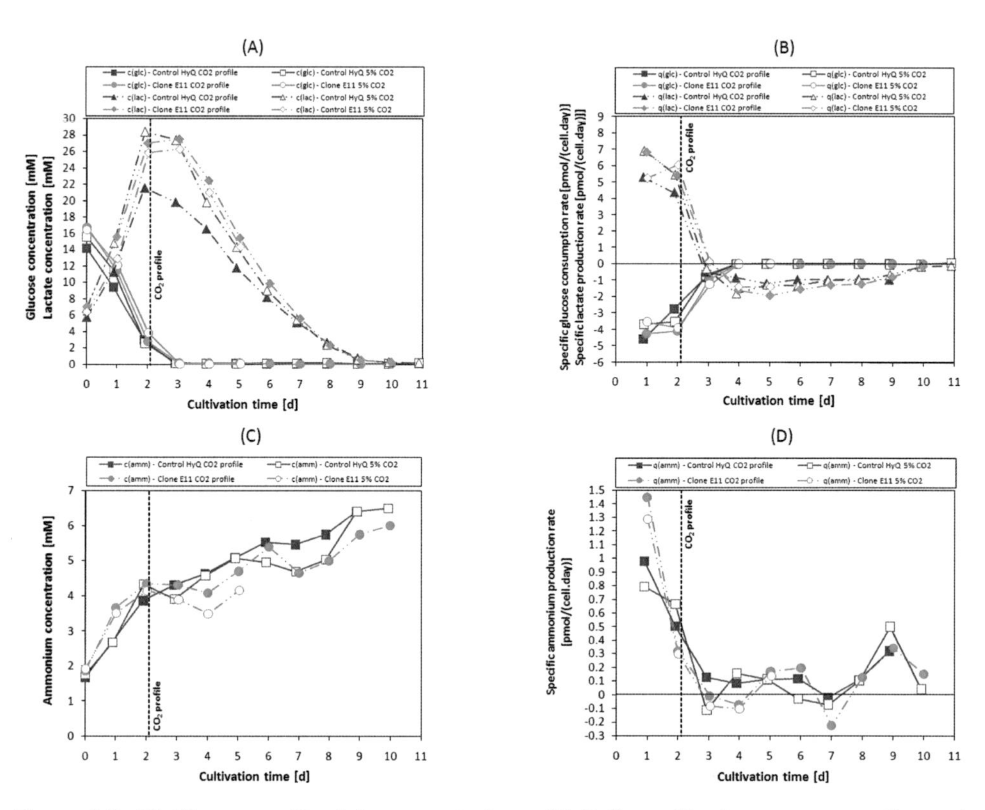

Figure 5.3: (A) Glucose and Lactate concentrations; (B) Cell specific glucose consumption and lactate production rates; (C) Ammonium concentrations; (D) Cell specific ammonium production rates.

differences were observed related to the cell lines or the CO_2 profiles.

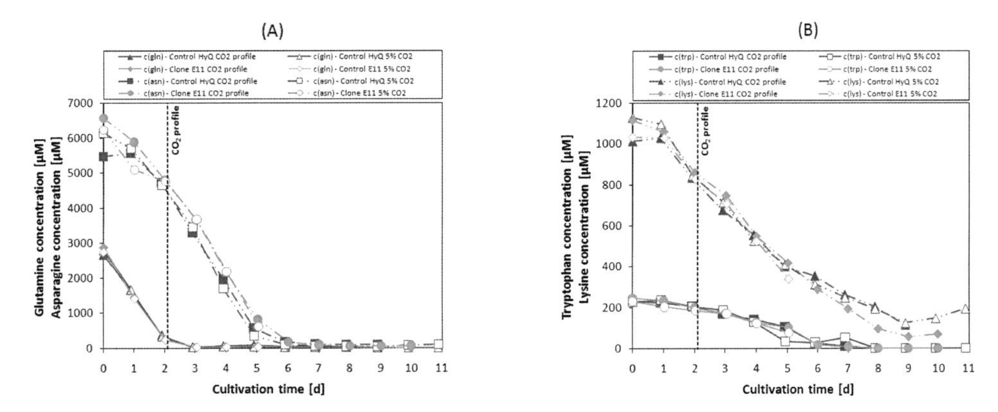

Figure 5.4: Amino acid concentrations. (A) Glutamine and asparagine concentrations; (B) Tryptophan and lysine concentrations

5.2 EVOLUTION OF INTRACELLULAR PH VALUE

The pseudo-null method was applied for flow cytometric measurements of the intracellular pH value during the cultivation process (Subsection 10.9.2). The culture extracellular pH was controlled to pH 7.2. The results for all cultivations are illustrated in the Fig. 5.5.

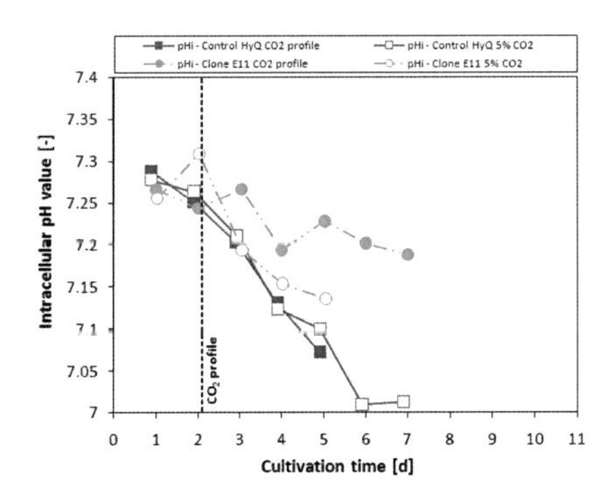

Figure 5.5: Intracellular pHi value during batch cultivations under extracellular pH controlling to 7.2 pH units.

For all four cultivations, the initial intracellular pH value was between pH 7.26 and pH 7.29. Subsequently, it decreases to pH values below the extracellular pH value in the bioreactor. Noteworthy is the slow decrease of the pH_i of the clone *E11* after the CO_2 profile started. When the profile reached its CO_2 maximum value of 25 % (cultivation day 6), the pH_i reached a value of 7.2. The clone *E11* reference culture (5 % CO_2) presented a pH_i value of 7.14 already at cultivation day 5. The pH_i of the control cell line cultures decrease linearly and it reached lower pH_i values than the ones registered for the hCAII-expressing cell line. At cultivation day 5 a difference of 0.16 pH units was calculated for the long term increased CO_2 cultures, while the difference between the 5 % CO_2 cultivations was minimal. The metabolic engineered cell line was able to maintain higher intracellular pH values during long term CO_2 increase cultivation with controlled extracellular pH value.

5.3 CELL CYCLE DISTRIBUTION

The percentage of cells in *G0G1* and *S* phases for batch cultures of both control cell line and hCAII-transfected cell line at controlled 5 % CO_2 and CO_2 profile are illustrated in the Fig. 5.6.

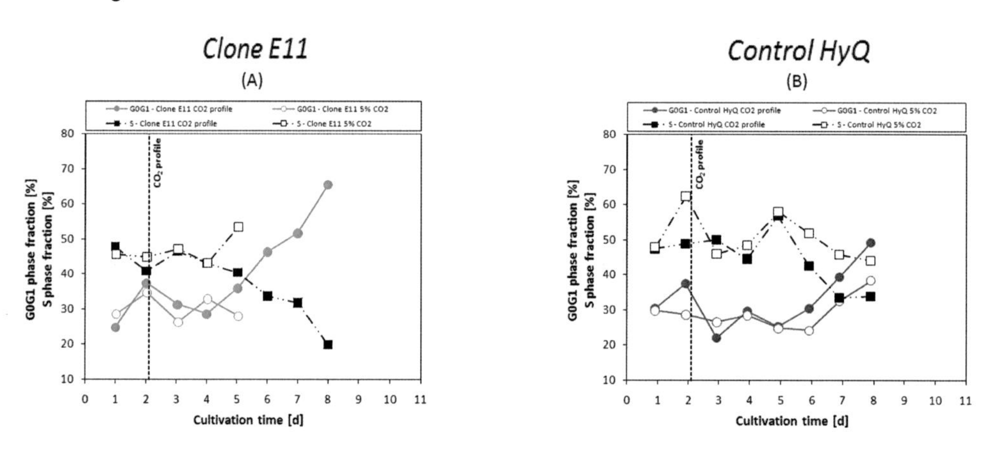

Figure 5.6: Changes in the fraction of cells in G0G1 and S phases during batch cultures of clone E11 (A) and Control HyQ (B) at controlled 5 % CO_2 and CO_2 profile

For both cell lines and cultivation types, the proportion of cells in the G0G1 phase started between 25 and 30 % whereas the fraction of cells in the S phase started at about 47 %. The amount of cells in the S phase maintained very high during all exponential growth phase (between 40 and 50 %), showing signs of cell activity related to active DNA synthesis. On the hCAII-expressing cell line cultured with a CO_2 profile, a progression of the cell cycle towards the increase of the percentage of cells in the G0G1 phase occurs already after cultivation day 4, reaching 65 % four days later. The same behavior is observed for the control cell line, starting one day later, though the degree of G0G1-phase arrest differed (49 % at day 8). Antiparallel, the fraction of cells on the S phase decreased reaching 20 % for clone E11 as opposed to 34 % for Control HyQ. In the cultivation of the control cell line at controlled 5 % CO_2, the same tendency was observed but the entry on the progress is slower. This result suggests a higher degree of cell proliferation suppression at increased CO_2 levels. No considerable alterations or significant tendencies were observed for the fraction of cells in the G2M phase (data not shown). The existence of an increased CO_2 profile from 5 to 25 % CO_2, trigger the cells to enter earlier in the G0G1 phase and the increase of the fraction of cells in this phase was faster for the hCAII-expressing cell line.

5.4 CARBONIC ANHYDRASE II EXPRESSION DURING CULTIVATION

During the cultivation, samples of every second day were taken for western blot analysis of the human carbonic anhydrase expression. The results are presented in Fig. 5.7.

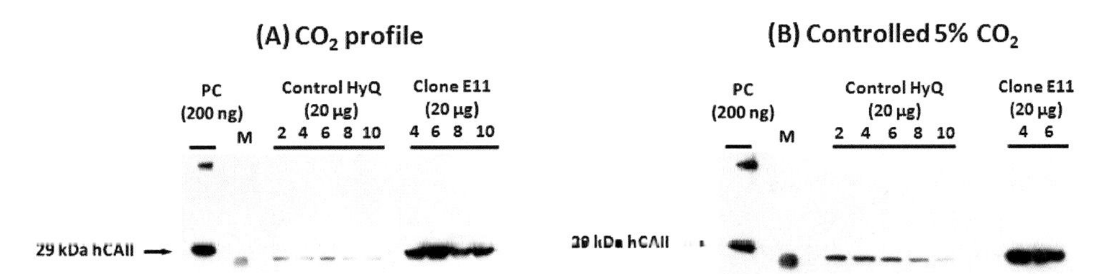

Figure 5.7: Immunoblot of hCAII in cell lines stably transfected with hCAII gene and in non transfected cells. (A) Immunoblot of samples from the CO2 profile cultivation. (B) Immunoblot of samples from the 5 % CO2 controlled cultivations. The numbers indicate cultivation days. Positive Control (PC): carbonic anhydrase isozyme II from human erythrocytes (Sigma-Aldrich); Protein marker (M): SeeBlue® Plus2 pre-stained standard (Invitrogen).

The detected carbonic anhydrase II had the same 29 kDa size as the positive control. The transfected cells showed a stable hCAII expression during the cultivation as demonstrated by the intensity of the bands. A lower expression of the hamster CAII was also detected in this immunoblot. The CAII antibody is polyclonal and therefore recognize multiple epitopes. It was proven that this antibody additionally reacts with mouse CAII (Abcam website information, ab6621). As carbonic anhydrases are highly conserved across mammals it might possibly be a cross reaction with the hamster carbonic anhydrase II. The detection of carbonic anhydrase II on the marker was already discussed on Subsection 3.1.3.

5.5 DISCUSSION

Only limited investigation is published where cells were submitted to increase continuous pCO_2 levels. JOCKWER (2008) performed fed-batch cultivations with the CHO-MUC1 cell line. Basically, there is limited online pCO_2 data during cultivations and if this information exists, the experiments were done in cell-free medium such as the ones from MOSTAFA AND GU (2003). With the increase of cultivation volume for recombinant proteins production is necessary a characterization of the effect of CO_2 on cell growth in similar conditions as it occurs in reality. For this effect the recombinant hCAII cell line was characterized under continuous increasing CO_2 levels in comparison to constant physiological levels. This parameter was separated from the hydrostatic pressure that may arise in large scale bioreactors.

The increased pCO_2 levels provoked a stronger suppression of cell growth for the hCAII expressing cell line. Thereby lower cell densities or growth rates were observed for the clone E11 than for the control culture but no significant alteration of cells metabolism was noted. The results of the different experiments related to the productivity showed that the cell lines, wether control or hCAII over-expressing cell line, react differently to diverse CO_2 conditions concerning productivity. In batch cultures, the cells growing in CO_2 profile presented higher recombinant protein productivity which was higher for hCAII cell line in the beginning of the culture and then similar for both cell lines after cell entering in the stationary growth phase. This effect was already observed by JOCKWER (2008) in a fed-batch process also with a controlled CO_2 profile with additional overpressure of 750 mbar for CHO-MUC1 cell line. No discussion of this result was done.

Since the human carbonic anhydrase II (hCAII) speeds the cytosolic CO_2 hydration equilibrium, under a large CO_2 burden it is expected that the transfected cells are able to speed up the re-alkalinization of its cytoplasm. This effect can only be evaluated at pH_i and cell cycle level. At an externally controlled pH, pH_i decreases with the specific growth rate. It is well known that at the beginning of a cultivation, pH_i is higher than pH_e and with increase cells metabolism, a decrease on pH_i is verified caused by lactate, ammonium and carbon dioxide accumulation in medium [WU *et al.*, 1993]. The hCAII metabolic engineered cell line presented more alkaline pH_i values than its control cell line during long term CO_2 increase cultivation with controlled extracellular pH value. Moreover, the pH_i was more alkaline for the cultivation under CO_2 profile indicating a positive action of the hCAII in respect to efficient re-alkalinization of cytoplasm. The existence of an increased CO_2 profile, triggered the cells to enter earlier in the G0G1 phase and the increase of the fraction of cells in this phase was earlier for the hCAII-expressing cell line.

During cell growth, there is a large percentage of cells in S phase. Toward the end of a batch culture more and more cells accumulate in the G1 phase mostly on the basis of suboptimal culture conditions, where lack of nutrients results in decreased biosynthetic activity, thereby reducing cell cycle progression. Since this is an idle state to the persistence of sub-optimal culture conditions, this phase is referred to as G0 phase. It therefore holds together both G0G1 phases, as they describe the same physiological state of the cell: The cell can not conclude the synthesis of internal components and is locked in the restricted point before the beginning of S phase [ALBERTS, 2002]. Some scientists describe relations between pH_i and cell cycle phases. Hence, it is generally an alkaline pH measurement in the cytosol when the cells are in a metabolically active phase of the cell cycle (S and G2 phase) [WELSH AND AL-RUBEAI, 1996]. Moreover, the pH_i promotes entry into the G2 phase of the S phase, and is thus partly responsible for the process of cell division [PUTNEY 2003].

By increasing the pCO_2 levels in a controlled manner, an efficient control of pH_i at alkaline levels, causing a cell cycle arrest in G0G1-phase and an increase in cell-specific productivity could be seen. The additional increase in productivity levels may be because more cells are in the G1 cycle phase, which has been reported to be the more productive phase in the cell cycle [Oh *et al.*, 1996]. The increase in specific productivity by arrest of cells in the G0G1-Phase is described in the literature. A 3-fold higher specific product formation was observed for example in the use of AMP (adenosine monophosphate), since the cells were arrested during the exponential growth phase and thus passed into a long stationary phase [CARVALHAL, 2003]. In the present work, the increase in specific productivity was stronger for the hCAII cell line than for its control only during the first 5 cultivations days. After that the increase of productivity was similar for both cell lines independent of pH_i value or cell cycle arrest. An interesting fact is that the final product concentration was 11 % higher and that overall cultivation was 1 day longer for clone cell line indicating an advantage of hCAII expression over the control cell line under pCO_2 increasing levels.

Overall, these results give insights that the hCAII recombinant cell line submitted to a CO_2 profile, has the lowest pH_i variations and also the more alkaline pH_i in the cultural history. Moreover, an increase level of cells in G0G1 phase after day 4 can be observed, representing a successful cell cycle arrest. The prolonged viable culture period, together with the effect of increasing specific productivity during the stationary phase of growth led to the highest end product concentration in this series of cultivations.

An important question, which arises with the transfection and the long-term cultivation of the genetically changed cells, is the stability of the integrated foreign DNA. The western blot analysis of the clone E11 cultivated in batch cultures showed that the hCAII gene expression remain stable over cultivation. This indicates stable transfection of this gene. Additionally, the endogenous CAII (hamster CAII) was also detected although the antibody should only react with human and mouse. This detection was also observed by LI and co-workers (2002) for AP1 cell, which are CHO cells that possess endogenous CAII. Results of detection for immunoblot with the same antibody for protein extracts and also immunoprecipitation samples. No discussion of the results was made. This antibody is a polyclonal antibody and the epitope is conserved among species, so it is most likely that its sequence is very similar with the mouse or human sequences and therefore it was also detected.

CHAPTER 6

RECOMBINANT PROTEIN GLYCOSYLATION

With the creation of metabolic genetic modified cell lines some questions arise concerning its effect on the product glycosylation. Furthermore, process development strategies and culture conditions can strongly modify culture performance and also alter protein glycosylation patterns. In this Chapter is reported the influence of the human carbonic anhydrase II expression and its cultivation under different controlled CO_2 profiles on a recombinant protein glycosylation.

6.1 INFLUENCE OF HCAII-EXPRESSION ON GLYCOSYLATION

The glycosylation profile of the human carbonic anhydrase II expressing cell line was compared with the one obtained from the original cell line. Samples were collected from the parallel shaker cultivation (Section 3.2) and the glycosylation profile of the recombinant protein was analyzed as described in Subsection 10.8.8. The relative abundances of complex glycans and mannose are shown in Fig.6.1.

Analyzing the relative content of complex glycans, agalactosylated (G0) and monogalactosylated (G1) structures were the most abundant. In the Clone E11 and Control

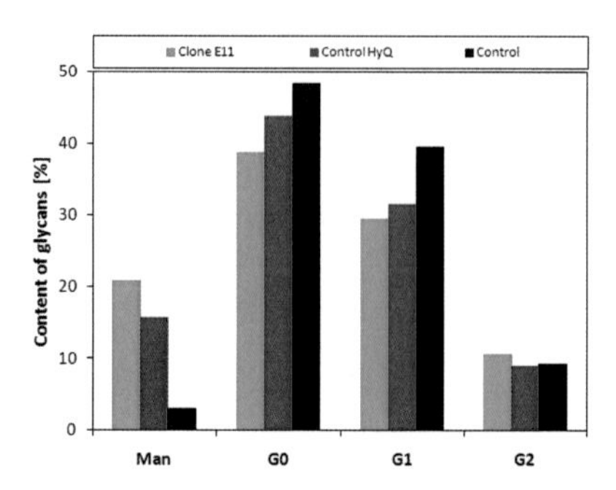

Figure 6.1: Comparison of the glycosylation profile of the recombinant protein produced by the hCAII-expressing cell line and the one produced by the original cell line, either Control HyQ or Control.

HyQ, complex glycans without terminal galactose (G0) represented 38.7 and 43.8 % of the total glycans, respectively, compared to 48.4 % for Control. While Control contained 48.7 % of galactosylated structures (G1 plus G2), the Clone E11 and Control HyQ showed a lower relative content of those structures (40.1 and 39.5 %, respectively). Opposed to that, an increase of the mannose structures is to be noted for the hCAII-expressing cell line (20.8 %) and Control HyQ (15.6 %) compared to the original cell line (3 %). Only minor differences are noted between the Clone E11 and Control HyQ but an undesired loss of product quality related to increase of mannose structures was observed in comparison with the Control.

6.2 INFLUENCE OF LONG TERM CO_2 INCREASE ON GLYCOSYLATION

Samples taken at the end of the bioreactor cultures at different controlled CO_2 levels of both clone E11 and Control HyQ (Section 5.1) were analyzed and the results are shown in Fig. 6.2.

The most abundant structures were G0 and G1. The relative amount of digalactosylated (G2) and mannose structures were below 15 and 10 % respectively of total glycosylation for all cultivations. The total amount of galactosylated structures showed a difference of 10 % between clone E11 cultivated at increased CO_2 levels and the Control HyQ cultivated at 5 % CO_2. These difference lies on 7 % between the control

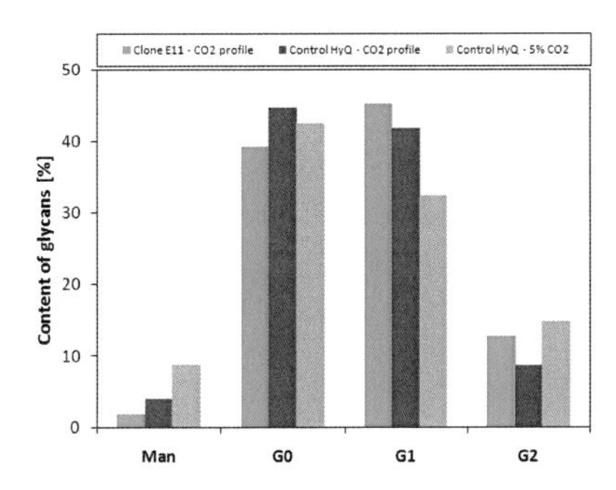

Figure 6.2: Comparison of the glycosylation profile of the recombinant protein produced by the cells under increased CO_2 profile with the one produced at controlled 5 % CO_2.

cell line at different increased CO_2 profiles. It can be speculated that the cultures with increased CO_2 levels delivered a slightly better glycosylation profile when looked at the mannose and G1 structures relative content. In general, no significant difference was observed in the glycosylation profile of the recombinant protein in all cultures.

6.3 DISCUSSION OF THE RESULTS

Given the importance of glycosylation for the functional properties of recombinant proteins and the necessity of a consistent production process for consistent product quality, the glycosylation pattern of the recombinant protein produced was investigated in more detail.

As G0 structures in antibodies correlate with various pathologies and that galactosylated structures predominate in normal individuals, one can infer that higher galactosylation should be desirable [SERRATO, 2007]. Complete glycosylation of recombinant proteins is usually associated with maximization of galactosylation and sialylation. Often these two processes are incomplete and this gives rise to considerable glycan structural variation. Only the secreted and completely post-translated recombinant protein is of pharmaceutical relevance [WALSH, 2009].

In industrial processes cultivation is frequently finished when cells viability reaches 40-60 %. With decrease of cells viability, the cells integrity is no longer ensured and cell lysis happens. This is accompanied by release of intracellular proteases into the culture broth, which can destroy the secreted recombinant protein. Additionally, intracellular recombinant protein is release, which processing may not be yet complete, *i.e.*, not completely folded and glycosylated. Under the condition that the shaking cultures were terminated at 40 % viability, the slightly decrease in G0 and G1 structures and the increase in the mannose structures after clone screening in HyQ media indicates that these procedure may had a negative impact on product quality. Though the minor differences in the glycosylation between hCAII transfected cell line and Control HyQ suggests that the hCAII expression doesn't affect product quality in batch culture.

In batch culture under different CO_2 profiles, also only minor differences were observed in product glycosylation between clone and control cell line or between high and low CO_2 levels. This indicates that the expression of hCAII and/or the CO_2 profile doesn't positive or negatively greatly affect the recombinant protein glycosylation.

The pattern of protein glycosylation is dependent on the expression of various glycosyltransferase enzymes that are present in the Golgi of the cell. Differences in the relative activity of these enzymes among species can account for significant variations in structure [BUTLER, 2005]. One of the parameters affecting enzyme activity is the extracellular pH value. Non-optimal pH conditions (<6.9 and >8.2) have also been shown to alter the pattern of glycosylation [ROTHMAN *et al.*, 1989; BORYS *et al.*, 1993]. During the all cultivation history the pH_e was controlled to 7.2 pH value and the pH_i ranged between 7.01 and 7.31, hence not outside of the non-optimal pH conditions that may negatively affect product glycosylation pattern. This is evidenced by the results obtained from the batch cultures under different CO_2 profiles.

ANDERSEN and co-workers (2000) suggested that site occupancy could vary with the growth state of cells and correlates with the fraction of cells in the G0G1 phase of the cell cycle. This suggests a mechanism by which glycosylation efficiency improves at a reduced rate of protein translation. In the present work, the hCAII transfected cell entered earlier in the G0G1 phase but no improvement of the glycosylation efficiency was noted.

The successful glycosylation of the product was investigated and it was shown that the clone screening nor the CO_2 profile had a strongt negative impact on product quality and that the recombinant protein expressed by the hCAII metabolic engineered cell line was completely glycosylated.

CHAPTER 7

PROTEOMIC ANALYSIS

The scientific aspect of proteomic involves the systematic study of the proteome. It comes from the English term "protein complement of the genome" and represents the whole expressed proteins of a cell (Wilkins, 1995). The first aim of the proteomic analysis is the comparison of the two cell lines, Clone E11 and Control HyQ, concerning their basis expression pattern. Furthermore, the shift of the proteome as a result of different pCO_2 conditions for the hCAII transfected cell line will be examined in more detail. The differential gel electrophoresis was applied for the investigation of the proteome.

7.1 COMPARISON OF THE EXPRESSION PATTERN OF CLONE E11 AND CONTROL HYQ

7.1.1 SELECTION OF SUITABLE SAMPLING POINTS

From the cultivations presented on Chapter 5, several samples were taken for proteomic analysis (Fig. 7.1). The first sample was taken at the time point were the cultivation conditions were comparable, i.e., before the start of the CO_2 profile (cultivation day 2), and in order to minimize the number of samples, the samples from the parallel cultivations of the same cell line were pooled (Control HyQ: sample A, Clone E11: sample D). The second sample was taken in order to investigate the influence of a high CO_2 level in the protein expression, therefore at cultivation day 4, with CO_2 13 %, the samples B and C were taken for Control HyQ and samples E and F for Clone E11.

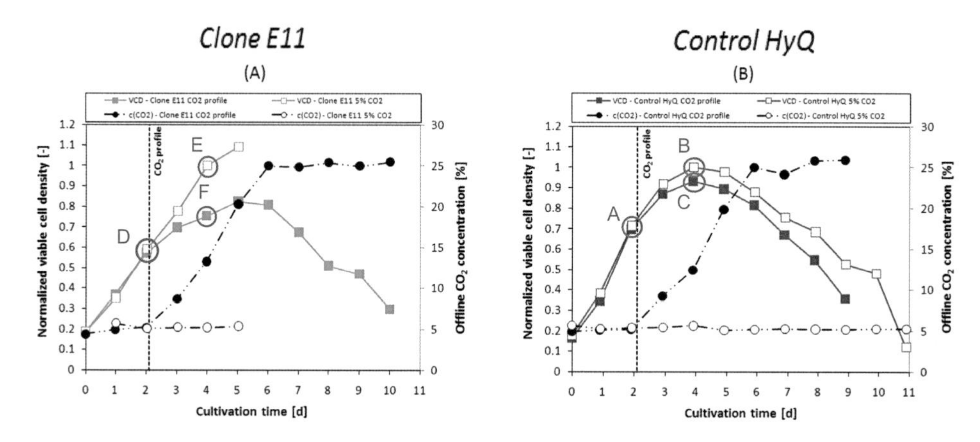

Figure 7.1: Sampling of the batch cultivations for proteomic analysis. (A) Clone E11 (B) Control HyQ. Samples A and D at day 2 (constant 5 % CO_2), samples B and E at day 4 (constant 5 % CO_2), samples C and F at day 4 (13 % CO_2).

7.1.2 ANALYSIS OF DIGE GELS

The analysis of the generated DIGE gels was done with the Delta 2D software. After the warping of the standard gel, spot detection was performed on the generated image fusion and 1038 spots were detected in the entire project.

To determine the number of proteins that significantly differ in expression between the six samples, an ANOVA (ANalysis Of VAriance) was performed with the implemented software TIGR Mev 4.0 Delta 2D. In contrast to the t-test, this form of analysis of variance can be used for comparison of several experimental conditions. The significance of a change in expression level is assessed here by comparing the relative spot volume. Is the variance of the spot volume between the groups under investigation greater than the variance within a group, the differences are not random, but rather resultant of an external influence. The ANOVA with a significance level of $\alpha < 0.01$ yielded 558 significantly regulated proteins. With a percentage of 54 % therefore changes occurs in more than half of the proteins. In order to identify differences in the expression between the individual samples, an heat map must be created. For this purpose, the relative spot volumes were logarithmically transformed and hierarchically clustered using the Euclidean distance by average linkage method. Through the hierarchical classification, the proteins and samples are arranged according to their similarity in expression pattern, which gives a general overview of the variance of the samples. Fig. 7.2 (A) shows the total heat-map, selected and extended sections from the heat map are shown in the Fig. 7.2 (B) and (C) and reviewed below.

The samples A to F were multiply clustered according to their expression pattern (Fig. 7.2 (A)). This decreases the variability in the expression profiles of samples with increasing subdivision. Accordingly, the samples of the two main clusters, samples A to C (Control HyQ) and samples D to F (Clone E11), show the largest expression differences. In most of the proteins from the different cell lines show a reciprocal expression. There is comparatively only few proteins that are similarly strong expressed in both cell lines. A section of the heat-map (Fig. 7.2 (B)) shows the proteins that have an increased expression in both cell lines (at cultivation day 4: sample B/C or E/F). In the course of a batch process there is a change in the culture conditions due to the decreasing substrate concentrations and the increasingly metabolic products. The cells have to adapt to changing conditions by changing their metabolism. Therefore, different expression patterns of the different sampling can be identified in both cell lines at different sampling times. However, there are large areas in the heat map where proteins contrast in their expression in the two cell lines (Fig. 7.2 (C). In this section, the proteins are strongly expressed in the Clone E11 (sample D to F) as in the Control HyQ (sample A to C).

The large differences in the expression of the two cell lines likely results from the genetic modification. To support this, a different growth and metabolism behavior was noticed in bioreactor cultures (subsection 5.1.1), which was not perceived in the parallel shaking flasks experiments when selecting the comparable clone for further experiments. This fact is elucidated and emphasized by the amount of differences of the clone E11 and its original cell line at the proteomic level. As there is no remarkable trend of proteins expressed at constant 5 % CO_2 or CO_2 profile, to have a better insight of the CO_2 effects at proteomic level only the clone E11 will be taken into consideration.

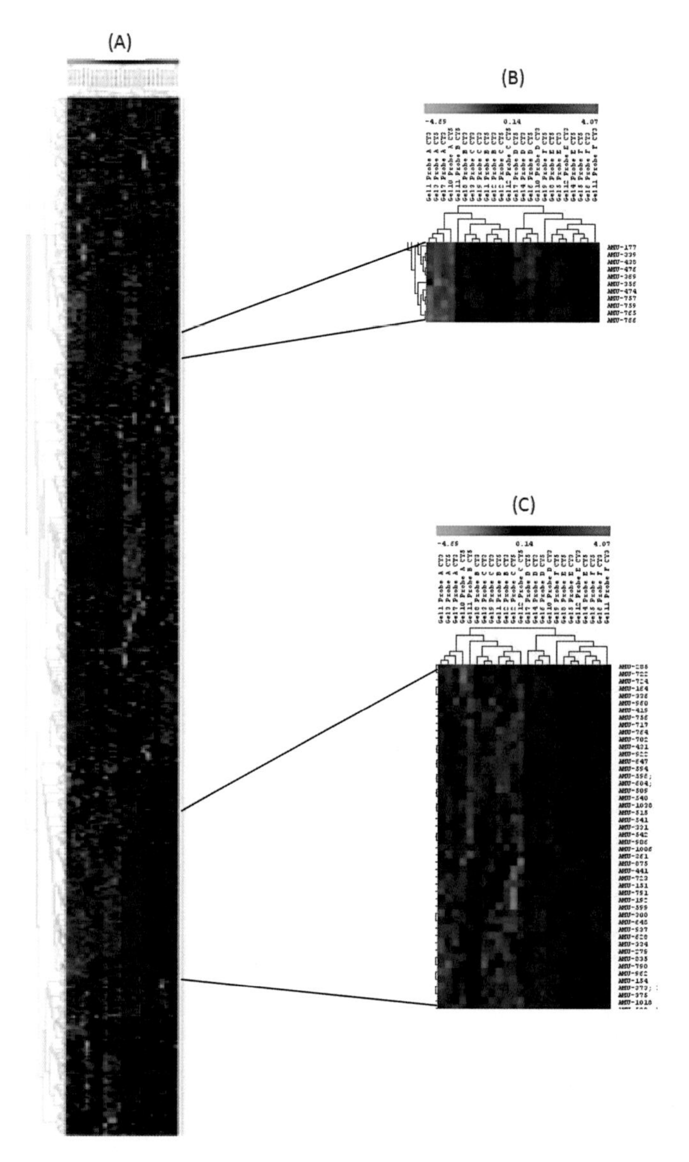

Figure 7.2: (A) Heat map of the samples A to F with hierarchically clustered. (B) Section of a heat-map of the samples A to F with hierarchical cluster analysis, illustrating examples of the differences in expression patterns on different sampling times. (C) Section of the heat-map of the samples A to F with hierarchical cluster analysis, illustrating the protein expression profiles of the differences between the two cell lines.

7.2 ANALYSIS OF THE EFFECTS OF INCREASED pCO_2 VALUES ON THE PROTEOME OF THE CLONE E11

A t-test was performed with a significance level of $\alpha < 0.01$ to determine the number of proteins, between samples E (day 4, 5 % CO_2) and F (day 4, 13 % CO_2) that the expression is significantly altered in the clone E11 cell line. The outcome of 82 proteins represents 8 % of the total 1038 proteins in the project. The proteins were identified using mass spectrometry.

7.2.1 PROTEIN IDENTIFICATION

2D gels were generated using the classical method (gels not shown). Thereby an amount of 450 μg total protein sample was applied and the separated proteins visualized using colloidal coomassie staining. With Delta 2D software, a fusion image of the gels was produced, which was aligned by warping with the fusion image of the DIGE gel, to be able to transfer the spot image (spotmaske) from the DIGE-project to the coomassie stained gels. Thereafter, the protein spots were picked from the gels, prepared and afterwards directly applied onto the MALDI target. Thirty eight protein spots were identified which corresponds to an identification rate of 46 %. However, some proteins were found in various spots, so in total there are only 30 different proteins. Because of the sparse database information, only 21 % of the proteins could be identified based on sequences of the hamster. The largest proportion of proteins was identified by comparison homology to the mouse (36.8 %) and rat (31.6 %). Moreover, for three proteins, the sequence information is based on human proteins (7.9 %), and even the sequence of a bovine protein.

The identified protein spots are shown in Figure 7.3 in a false color of a 2D-DIGE gels with the samples E and F. A detailed overview of the proteins with information on their molecular weight, pI, the calculated MOWSE-score, the sequence coverage and the relative spot volumes can be found in Appendix A. In subsequent chapters, are presented the classification of the identified proteins into functional categories and the interpretation of their expression profiles in the different samples.

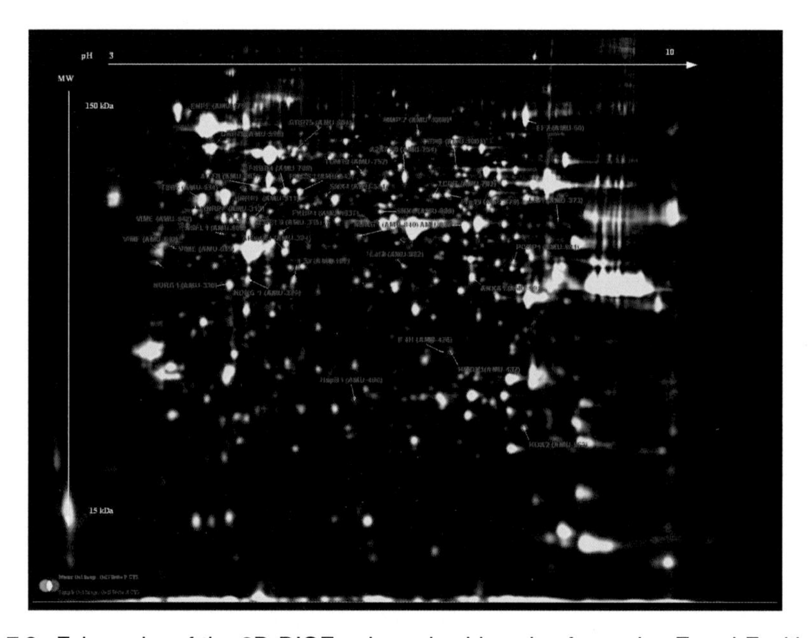

Figure 7.3: False color of the 2D-DIGE-polyacrylamide gels of samples E and F with the 37 identified proteins. The protein extract from sample E (5 % CO_2) was marked with Cy^{TM}5 and sample F (13.3 % CO_2) with Cy^{TM}3. The protein spots from the two samples that not differ in their relative intensities are presented as yellow spots.

7.2.2 CLASSIFICATION OF THE IDENTIFIED PROTEINS INTO FUNCTIONAL CATEGORIES

The identified proteins are grouped in the 9 functional categories: "stress proteins / chaperones", "structure proteins", "transport proteins", "biosynthesis proteins", "mRNA processing", "lipid metabolism", "redox control", "DNA replication" and "others". Proteins with unknown or different functions were grouped in the category "others". Figure 7.4 shows the proportions of the proteins of the above categories based on the total number of identified proteins.

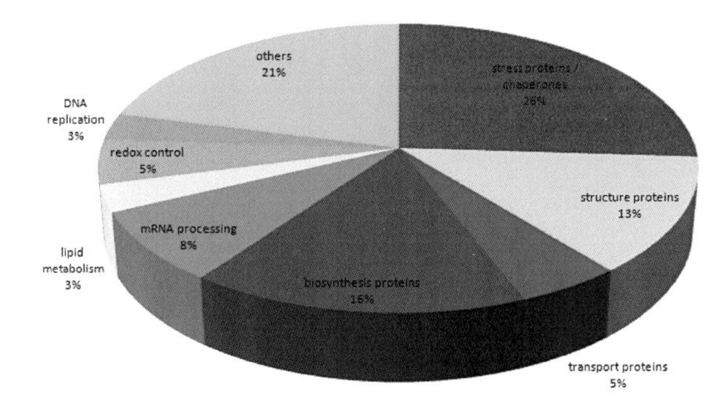

Figure 7.4: Contribution of individual functional groups on the total number of proteins identified.

As can be seen in Figure 7.4, the majority (26 %) of the identified proteins have a role in the stress response of the cells. The proteins in the category "others" with 21 % is the second highest percentage of the total amount of identified proteins. Then follows the categories "biosynthesis proteins" (16 %) and "structural proteins" (13 %) as well as components of the "mRNA processing" (8 %).

7.2.3 HIERARCHICAL CLUSTER ANALYSIS OF THE IDENTIFIED PROTEINS

A hierarchical cluster analysis was performed to clarify the expression profiles of the identified proteins in the samples E (clone E11, day 4, 5 % CO_2) and F (clone E11, day 4, 13 % CO_2). The heat map and the respective expression profiles are shown in Figure 7.5. The columns represent the mean values of the normalized spot volumes in which the determined standard deviation was inserted as horizontal bars.

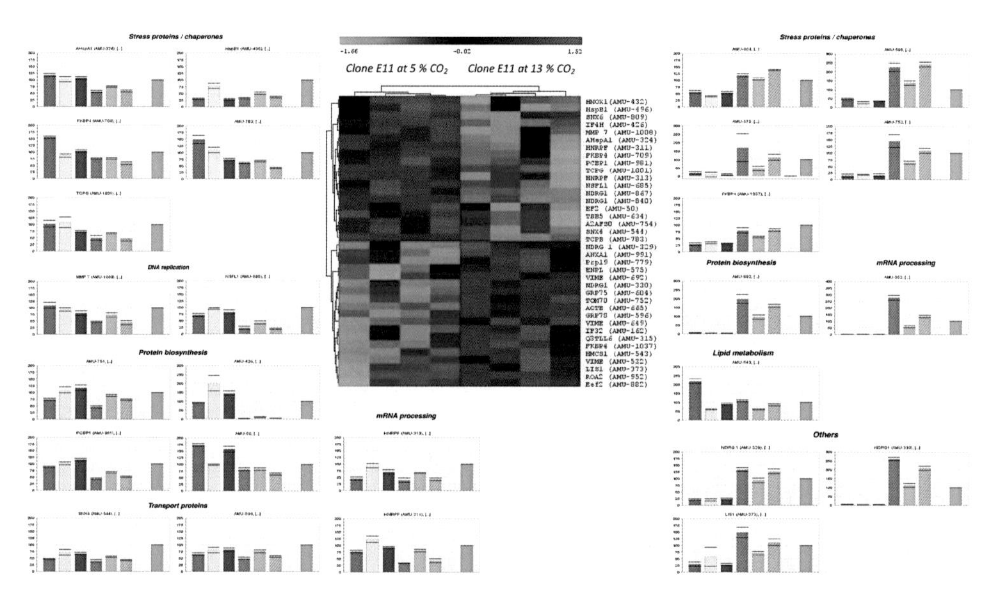

Figure 7.5: Heat map of hierarchical cluster analysis of the identified proteins for Clone E11 samples E (5 % CO_2) and F (13 % CO_2). Mean expression profiles of significant stress proteins in clusters. The colors of the columns: Sample A (Control, day 2, 5 % CO_2), B (Control, day 4, 5 % CO_2), C (Control, day 4, 12 % CO_2), D (Clone E11, day 2, 5 % CO_2), E (Clone E11, day 4, 5 % CO_2) and F (Clone E11, day 4, 13 % CO_2). The green column shows the average spot volume of the applied standards, which was set at 100 %.

Two clusters arise in the arrangement of the proteins according to their expression in the samples, which hereinafter correspond to the up-regulated protein expression in the culture with constant CO_2 (left part, marked in red), whereas marked in red in the right part are the proteins with increased expression in the culture with the CO_2 profile. Both clusters contain 19 proteins, however, due to multiplicity, remain 17 and 15 different proteins for the first and second cluster, respectively. Following will be explained the role of the selected proteins in the cells on the basis of the determined expression profiles.

Stress proteins and chaperones

The proteins AHspA1, HspB1, TCPB, TCPG and FKBP4 have been classified into the category of stress proteins and were up-regulated in the 5 % CO_2 cultivation. These proteins are almost exclusively cytosolic chaperones. The protein HSPB1 is a cytosolic heat shock protein of Hsp20 family as well as the T-complex proteins TCPB and TCPG, HSPB1 plays a role in the folding of the structural proteins actin and tubulin [LLORCA *et al.*, 2001]. The Activator of 90 kDa heat shock protein (AHspA1) and FKBP4 are among the co-chaperones. Both proteins stimulate the chaperone Hsp90 activity through their activity [FEIGE, 1996]. Chaperones are not only expressed under stress conditions, but also play a role under normal conditions in the *de novo* synthesis or transport proteins. Most of the identified stress proteins are involved in the conservation of structural proteins such as actin and tubulin. Therefore, their expression could be related to the growth of cells since at this time the culture under constant CO_2 conditions grew significantly better. This also explains the increased expression in the control HyQ cultures, which evidence a better growth then the clone E11 cultures.

In the sample F most of the stress proteins identified with higher expression are Glucose Regulated Proteins (GRP) of the endoplasmic reticulum. These include GRP75, GRP78 and Endoplasmin (ENPL, including: GRP94). In addition, the mitochondrial membrane protein TOM70 and the already mentioned co-chaperone FKBP4 fall into this category. The expression of GRPs is related to the Unfolded Protein Response (UPR) in the endoplasmic reticulum. The activation of the UPR is effected by factors other than the heat shock proteins. It is caused for example by glucose limitation, an accumulation of unfolded proteins or the blocking of the glycosylation. One consequence of the UPR is to reduce the overall protein biosynthesis procedures to prevent further accumulation of unfolded or misfolded proteins. The formation of chaperones is increased, which include the belonged identified representatives GRP75, GRP78 and

ENPL (GRP94). Its responsibilities include the correct folding or the initiation of the degradation of accumulated proteins [NI AND LEE, 2007]. The co-chaperone TOM70 is situated in the outer membrane of mitochondria and interacts with Hsp70 and Hsp90 [VOOS, 2003]. Through this interaction, a translocation of the newly formed proteins is accomplished through the membrane, making sure that the correct form is preserved during the process. For the previously mentioned FKBP4 an interaction was found with GRP78 [FEIGE *et al.*, 1996].

DNA replication

The components involved in DNA synthesis are also directly related to the growth of the cells. The identified mini-chromosome maintenance protein 7 (MMP7 or mcm7) assumes a role in the initiation of DNA replication. It forms a complex with proteins mcm4 and mcm6 and binds to DNA before starting the replication. There is evidence of helicase function of all three proteins [ISHIMI, 1997] and therefore it is assumed that they contribute to the opening of the DNA double-strand. The up-regulation of the MMP7 protein show a stronger growth of the constant CO_2 (sample E) compared to the CO_2 profile (sample F) culture. This is also supported by the increased expression of the protein NSFL1 (N-ethylmeleimide-sensitive factor cofactor 47) which is a peripheral membrane protein that is involved in membrane fusion [WICKNER AND W. SCHEKMAN, 2008]. Both proteins are also stronger expressed in the control HyQ.

Biosynthesis proteins, transport proteins, mRNA processing

Three of the 19 identified up-regulated proteins in the 5 % CO_2 culture play a role in the protein synthesis. These include the eukaryotic translation initiation factor 4H (IF4H), the poly(rC) binding protein 1 (PCBP1), the elongation factor 2 (EF2) and the cytoplasmic seryl-aminoacyl-tRNA synthetase (A2AFS0). IF4H and PCBP1 play a role in the initiation of translation of mRNA products in the amino acid sequence. The initiation can be done in two ways and PCPB1 does have a particular task. In the first mechanism the initiation factors (including IF4H) bind to the 5'Cap of mRNA and thus facilitate binding of ribosomes. PCBP1 is used for the alternative mechanism, where it binds to the internal ribosome entry segment (IRES) region on the mRNA. This route takes, in addition, to the cap-dependent initiation [DOBBYN *et al.*, 2008]. In contrast, the seryl-tRNA synthase amioacyl, catalyzes the loading of the tRNA molecules for their specific amino acid. The up-regulation of these proteins in the sample E can be related to the stronger growth of the cell line, although at this cultivation time the clone E11 submitted to a CO_2 profile showed a higher productivity in respect to the recombinant antibody.

The identified biosynthesis proteins up-regulated in the CO_2 profile culture involve the fragment of the elongation factor 2 (Eef2), most probably a degradation product. The Eef2 fragment appears only in the samples of the clone E11 and it is stronger expressed in the samples D and F. This result may be associated with the stress reaction. A large amount of the fragment is detected in the samples, in which a high expression of GRP proteins was detected. Not correctly folded or modified proteins are degraded by the unsuccessful repair by chaperones. Therefore, the increased presence of these degradation products in the clone E11 would support the assumption that in this cells there is an accumulation of misfolded proteins that comes to their degradation.

The identified transport proteins sorting nexin 4 (SNX4) and sorting nexin 6 (SNX6) also indicate an increased protein synthesis in the sample E. Similarly at level of mRNA processing one major activity of the identified proteins is seen in the 5 % constant CO_2 clone E11 culture. The mRNA processing occurs in eukaryotes, since the gene sequences are interrupted by non-coding DNA sequences and therefore, the formed mRNA products during the transcription are not made directly accessible to the translational apparatus and must first be processed. This process is called splicing. Due to the different combinations of exons (alternative splicing) different proteins can be formed in the subsequent translation from a mRNA product [BERG *et al.*, 2003]. Among others, the heterogeneous nuclear ribonucleoprotein F (hnRPF) makes part of the splicing apparatus.

Another example of degradation products in cell line E11 is the identified Heterogeneous nuclear ribonucleoprotein A1/B2 (hnRPPA1 or ROA2), the sole representative of the increased expression of the mRNA processing proteins in the CO_2 profile cultivation (sample F). It is involved in the formation of ribonucleosome and is taking a role in various processes such as transcription, mRNA splicing, the translation or the RNA transport. Examination of the false color gel (Fig 7.3) comparing the data from Appendix A it can be concluded that the protein may not still be in the intact form. Thus, a molecular weight of 36.0 kDa is indicated for this protein and in the gel, however, the identified spot is below the HSPB1 protein whose molecular weight is 23.5 kDa. A MOWSE score of 150 and a sequence coverage of 54 % was calculated for HSPB1 indicative of high probability that it is HSPB1. For ROA2 a MOWSE score of 96.5 and a sequence coverage of 23.5 % was determined, thus it is most likely to be a degradation product of the protein.

Lipid metabolism

The cytosolic hydroxymethylglutaryl-CoA synthetase (HMCS1) catalyzes the reaction of acetoacetyl-CoA and acetyl-CoA (pyruvate) to hydroxymethylglutaryl-CoA [BERG *et al.*, 2003] and is up-regulated in the sample F. This is the source material for the formation of cholesterol, which is in general important for cell growth as a component of the membrane and specifically play a role in ovary cells as a precursor of progesterone. Therefore, in principle, larger amounts of this protein are to be found in ovary cells [HEGARDT, 1999]. In this work, a stronger expression of HMCS1 was detected in both cell lines, clone E11 as well as in control HyQ, on the first sample point. One explanation could be the initiation of glucose limitation. At the first sampling time, the cells used pyruvate formed from glucose possibly besides the facilitation into the TCA cycle and also to the formation of cholesterol. On the second sampling point, however, the pyruvate may be incorporated into larger units for energy generation in the TCA cycle.

Others

The function of the protein NDRG1 (N-myc downstream-regulated gene 1 protein) is still not exactly clear. It probably does have a role in triggering the growth arrest and cell differentiation and possibly acts as a tumor suppressor protein [ELLEN *et al.*, 2008]. The expression of the protein can be enhanced by various stress factors, such as p53, Egr-1, HIF-1α, myc, AP1, which eventually leads to growth arrest [ELLEN *et al.*, 2008]. This stress factor induced NDRG1 protein shows much higher expression in the clone E11 cultures and is up-regulated in the sample F, whereas only a very low expression is detectable in the control cell line. No reasonable explanation can be done regarding the cell cycle since the phases profiles didn't differ that significantly for those samples.

The Platelet-Activating Factor Acetylhydrolase (LIS1) catalyzes the deacetylation of phospholipids. Under stress, it initiates the destruction of damaged lipids and thus contributing to maintaining the cell structure (Arai et al., 2002). The LIS1 protein shows a similar expression profile as the one for NDRG1.

7.3 DISCUSSION OF RESULTS

For the proteome analysis, the two cell lines were cultured under both 5 % constant and gradually increasing CO_2 conditions in a controlled system. The sampling occurred in all cultures at the same time, the first sample was taken before the start of the CO_2 profile and the second on the fourth cultivation day were the cells were exposed to the different CO_2 levels.

Through the hierarchical cluster analysis was first recognized that the two cell lines had very different protein expression patterns. This result correlates with the results of various works in which the comparison of genomes and proteom of recombinant cell lines generated from the same cell line was performed [NISSOM *et al*, 2006; PASCOE *et al.*, 2007; SETH *et al*, 2006]. It could be demonstrated that the introduction of a gene caused numerous changes in the gene expression which were related with the phenotypic differences between cell lines regarding growth patterns, productivity or metabolism. The hierarchical cluster analysis also permitted to identify different expression patterns on the various sampling points. This result is in accordance with expectations, since the cells adapt their protein expression in the course of the change in the conditions of the batch cultivation.

In order to assess the effects of the hCAII expression at high CO_2 levels, the focus was towards clone E11 at different CO_2 levels in the same cultivation stage. The main part of the identified proteins were stress proteins as well as protein synthesis and structural components of proteins. Similar results have also led to other publications, in which a proteome of Sp2/0 or CHO cells was performed [WINGENS, 2008; NISSOM *et al*, 2006; PASCOE *et al*, 2007]. The reason probably lies in the abundance of cytoplasmic proteins on the basis of a more successful recovery, isolation and identification of these proteins. Proteins from the membrane, organelles or nucleus, however, are usually under represented. This is a disadvantage of the non-fractionated 2D gel electrophoresis [ABDATZADE-BAVIL, 2004].

Mainly cytosolic stress proteins were up-regulated in the 5 % CO_2 clone E11 culture. These proteins play a role in stabilizing the actin and tubulin cytoskeleton elements and some of the components are involved in protein synthesis. Based on this expression profile an enhanced cell growth of this culture could be appointed. While both cultures shows under constant CO_2 levels exponential growth until the second sample, after which the profile culture entered in the stationary phase. The increased formation of the recombinant antibody under the conditions of the CO_2 profile can not be shown using the expression profiles. However, the cells do not only do protein synthesis for the production of recombinant antibodies, but also for growth and metabolism. Therefore, the expression profiles presented the results of the productivity of recombinant antibody does not necessarily contradict.

Mainly glucose regulated proteins (GRP) of the endoplasmic reticulum were up-regulated indicating an increased unfolded protein response (UPR) at elevated CO_2 levels. It is unlikely that glucose limitation triggered the stress response in the CO_2 profile culture, since both cultures showed a similar substrate consumption. Peculiar is the fact that the proteins involved in this section are much lower expressed in the original cell line, indicating a phenomena related to the transfected cell line. The stress response in clone E11 cell line is therefore probably due the hCAII insertion with lateral effects.

It could be that there is an accumulation of the products formed, since the cells for example, fail to modify the overexpressed proteins in a thoroughly tolerable period. One indication of this can also be deduced from the results of Western blot (Figure 5.7. Here, by comparison with the positive control a quantity of 200 ng CAII was estimated at an applied amount of 20 μg of total protein for the clone E11. That would mean that the produced hCAII quantity is itself a big percentage of the total amount of protein in the cell line E11. In the Control HyQ (samples B and C), however, the stress proteins GRP75, GRP78 and GRP94 are only slightly up-regulated under elevated CO_2 conditions compared with the constant CO_2 culture. Hence, it can not be said whether there is a correlation between the expression of stress proteins and the increased CO_2 level but rather related to accumulation of the overexpressed proteins, such as hCAII. The depressed growth of the clone E11 due to a stronger stress response could also be demonstrated at the protein level by regarding the expression profiles of the control cell line.

It is well documented that cell characteristics can change when cell lines are cultivated for extended periods [WENGER *et al.*, 2004]. Cell lines that have been over-subcultured can experience phenotypic as well as genotypic changes (genetic drift). The increased unfolded protein response (UPR) could also be related to the clone screening method used in this project. During this step, cells are maintained in culture for long periods of times and the originated cell line might no longer considered a true model of the original source material [WENGER *et al.*, 2004]. However, the all selection procedure had a duration of about 1.5 month, which is considerably low.

It turned out to be very difficult to bring the expression of the identified proteins in the context of the CO_2 conditions. A major problem in the interpretation of the results was the fact that no proteins were identified related to the direct or indirect effects of CO_2. It is further assumed that the identified changes are most probably related to differences in cell density and metabolism. To identify the proteins involved in such side effects, a comparison between the cultures with the first sample would have to be made. Since the samples of comparative cultures were pooled for the purpose of reducing the number of samples, this comparison could not be performed. Altogether, only minor expression differences are perceptible between samples E and F, where the effects of elevated CO_2 levels are only indirectly associated with the decreased growth and protein synthesis components.

CHAPTER 8

CONCLUSIONS

With the creation of a cell line expressing active human carbonic anhydrase II enzyme, a stable system exists that permits the investigation of the CO_2 effects on mammalian cell growth and metabolism and product quality. The following consolidated findings of this project are:

Generation of a hCAII expressing cell line

- short single clone screening procedure (ca. 1 month) with no effect on cell growth, metabolism or glycosylation in batch cultures.
- successful production of active hCAII enzyme through linear DNA transfection of hCAII gene in CHO cell line.

Physiologic effect of CAII on pH_i

- demonstration of the feasibility of using flow cytometric and off-line gas analysis to monitor the steady-state as well as rapid pH_i changes in mammalian cells.
- observation of better re-alkalinization of cytoplasm associated to higher initial recovery rates after CO_2 acid load for hCAII expressing cell line
- indication that the pH_i fine-tuning at physiological pH_i might be performed by the Na^+-dependent HCO_3^--Cl^- antiport (NCBE) and the Na^+-independent HCO_3^--Cl^- antiport (AE) and not by the Na^+-H^+ exchanger (NHE1).

- indication that the presence of hCAII might have a crucial role in the short-term pH recovery in industrial mammalian cell culture conditions.

Long-term CO_2 increase on cells

- constrainment of the cell growth of the hCAII expressing cell line by the existence of a CO_2 profile.
- observation of the lowest pH_i variations and also the more alkaline pH_i in the presence of a CO_2 profile for the hCAII expressing cell line.
- triggering of the entrance in the G0G1 phase of the cell cycle with associated higher specific productivity for both clone and control cell line in the presence of the CO_2 profile.

Recombinant protein glycosylation

- minor negative impact on glycosylation due to the clone screening procedure and not due to the hCAII expression.
- no negative effect on glycosylation as a result of the CO_2 profile and the hCAII expression.

Proteomic analysis

- observation of significant different protein expression profiles of the hCAII expressing cell line compared to its control cell line indicating that hCAII transfection provoked changes in cells genome.
- stronger relation of the protein expression to hCAII overexpression than the potential benefits of the hCAII expression in the presence of the CO_2 profile.

Chapter 9

Future perspectives

The results of this project show the potential of the hCAII expressing cell line for the investigation of the effects of CO_2 at short and long term acidification for bioprocess optimization.

In order to exclude negative interferences of the transfection of genes, an isogenic cell line should be created with, for example, the Flp-In^{TM} (Invitrogen) or recombinant cassette system to have only single integration of hCAII and not multiple copies. Furthermore, another elegant and, at the same time, industrially interesting variant of the increase of the gene expression is the introduction of the hCAII gene into the DHFR-deficient CHO cell line and make gene copy number amplification with methotrexat [Wurm *et al*, 1986]. Further exists the possibility of switching "on" or "off" the expression of the hCAII-gene by the use of a constitutive promoter.

The suggestion that NCBE and AE are more deeply involved in the pH_i recovery at physiological pH should be investigated more intensively through inhibition of the transporters. Further analysis should also be performed in order to identify if there is really association of these exchangers with hCAII in form a of complex metabolon.

With the realization of this project an increased need to produce at scale data of the increase of CO_2 level on a large-scale mammalian cell cultivation was identified. An interesting experiment could be the measurement of the pCO_2 level at several heights of a large scale bioreactor to create accurate data of the pCO_2 profile along the bioreactor's height generated by high hydrostatic pressure effects. The hCAII expressing cell line could then be cultivated in such a simulated environment and the effects of a possible wide pCO_2 profile due to hydrostatic pressure and the high mixing times on the cell's could be more accurate and deeply investigated. The potential advantage of the expression of the hCAII under this conditions could be explored.

In order to take advantage of the long term alkaline pH_i and the entrance of the cell in the G0G1 phase of the cell cycle on the increase of the productivity, an "on-off" mechanism should be designed where hCAII expression could be regulated during the different cultivation phases. More tests should be done maybe in a continuous culture in order to identify a stronger correlation between pH_i-cell cycle-productivity and to plan a better strategy for running such a programmed cultivation.

CHAPTER 10

MATERIALS AND METHODS

If not otherwise mentioned, the terms centrifugation and spin down for the generation of supernatants/pellets are understood as a centrifugation at 200x *g* for 5 min.

10.1 MOLECULAR BIOLOGICAL METHODS

When foreign DNA is introduced into the cell, it will readily recombine with any parts of the chromosome bearing a significant homology. To genetically prepare clones of CHO cells, a three step approach is taken. First, the gene of interest (GOI) is cloned into an appropriate mammalian expression vector. Second, the resulting plasmid is prepared in *E. coli*, and finally, the construct-bearing plasmid is introduced into the CHO cells, where the desired DNA is incorporated in the genome.

10.1.1 VECTORS

pIRES2-ZsGreen1 vector

In the present work, the commercial available vector pIRES2-ZsGreen1 (Clontech) was used. The vector map and multiple cloning site sequence are presented in Fig. 10.1.

The pIRES2-ZsGreen1 is a vector consisting of a 5283 bp vector. It allows stable and transient expression of foreign genes in many mammalian cell lines. The vector con-

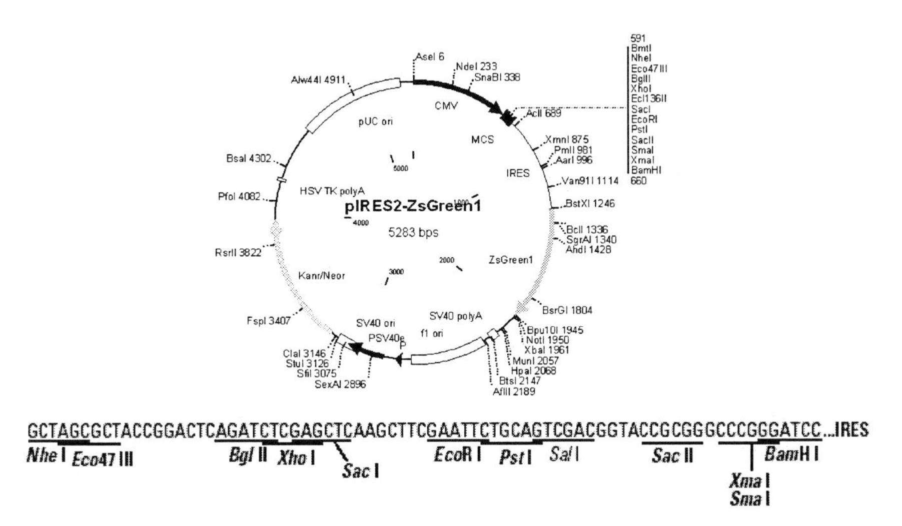

Figure 10.1: Map of the pIRES2-ZsGreen1 cloning vector with respective multiple cloning site (Clontech). Unique restriction sites are in bold

tains the internal ribosome entry site (IRES; JACKSON *et al.*; 1990, JANG *et al.*, 1990) of the encephalomyocarditis virus (ECMV) between a MCS and a region coding the *Zoanthus* sp. green fluorescent protein (ZsGreen; MATZ *et al.*, 1999). This design permits the GOI, cloned into the multiple cloning site (MCS), to be co-translated with the ZsGreen1 gene from a single bicistronic messenger ribonucleic acid (mRNA). SV40 polyadenylation signals downstream of the ZsGreen1 gene direct proper processing of the 3' end of the bicistronic mRNA. The vector backbone also contains a SV40 origin of replication in mammalian cells expressing the SV40 T antigen. A neomycin-resistance cassette (Neo^r), consisting of the SV40 early promoter, the neomycin/kanamycin resistance gene of Tn5, and polyadenylation signals from the herpes simplex virus thymidine kinase (HSV TK) gene, allows stably transfected eukaryotic cells to be selected using G418. A bacterial promoter upstream of this cassette expresses kanamycin resistance in *E. coli*. The pIRES2-ZsGreen1 vector also provides a pUC origin of replication for propagation in *E. coli* and a f1 origin for single-stranded DNA production (Clontech).

ZsGreen1 is a human codon-optimized ZsGreen variant that encodes a green fluorescent protein [HAAS *et al.*, 1996]. The protein has a excitation maximum at 493 nm and a emission maximum at 505 nm and, therefore, it is suitable for flow cytometric analysis.

pJRC36 vector

The mammalian expression vector for hCAII gene (NCBI Ref. Seq.: NM000067) was kindly provided by Dr. Casey (University of Alberta, Canada). It has a size of 4807 bp and it confers ampicillin resistance (Amp^r) in *E. coli*. In Fig. 10.2 is the vector map of pJRC36 presented. The vector construction is described by STERLING AND CASEY [1999]

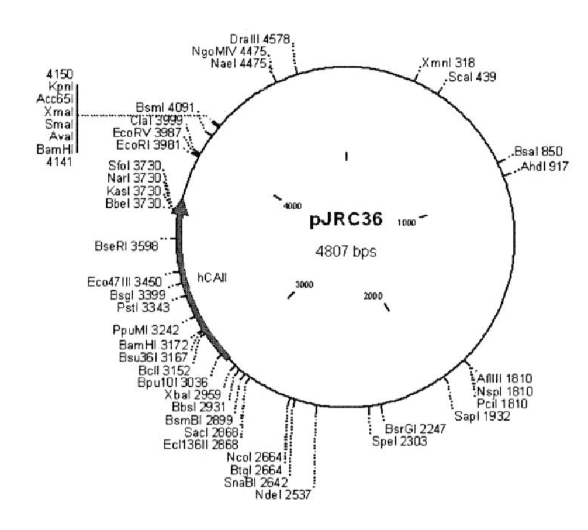

Figure 10.2: Schema of the cloning vector pJRC36

10.1.2 STRATEGY FOR VECTOR CONSTRUCTION

It was intended to express the gene encoding for hCAII and a fluorescent reporter protein under the same high-level expression promoter ($P_{\text{CMV IE}}$) in a single bicistronic mRNA. This was achieved by restriction (SacI and EcoRI) of the hCAII gene from the pJRC36 vector and insertion into the MCS of the commercial available pIRES2-ZsGreen1 vector (Subsection 10.1.1), previously restricted with the same enzymes. The constructed mammalian expression vector was used for the stable transfection of CHO cells using the nucleofection technique (Section 10.3). Selection of hCAII-expressing cells was done with G418 and single clones were isolated by limited dilution (Subsection 10.6.2). The molecular biological methods necessary to construct this expression vector are described in the following Subsections.

Both pIRES2-ZsGreen1 and pJRC36 vectors were digested with the restriction enzymes SacI and EcoRI. This restriction generated fragments with sticky-ends. Both fragments were purified by agarose gel electrophoresis and gel extraction before ligation. The 1109 bp insert containing the hCAII gene was cloned into the 5274 bp pIRES2-ZsGreen2 backbone resulting in a 6383 bp vector (Fig. 10.3). The construct contains 126 bp 5′-untranslated and 198 bp of 3′-untranslated. Control restriction to inspect existence and orientation of the hCAII insert was performed with BamHI resulting in two fragments (5543 bp and 840 bp).

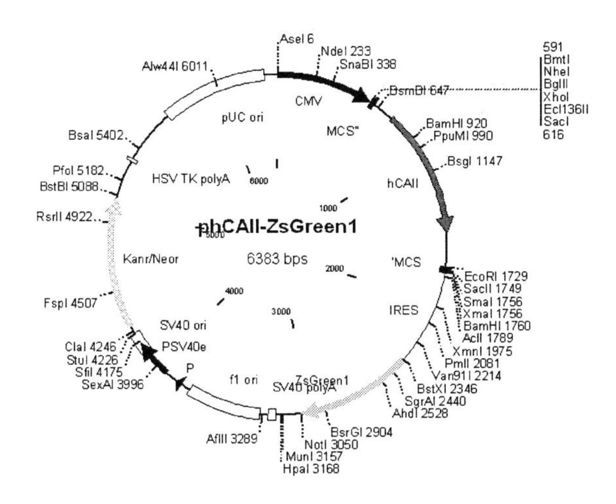

Figure 10.3: Schema of the cloning vector phCAII-ZsGreen1

10.1.3 PLASMID DNA RESTRICTION

The restriction of plasmid DNA was carried out according to manufacturers instructions for the particular enzymes used. The enzymes used in this work are presented in Table 10.1 and the reagents and respective amounts used are indicated in Table 10.2.

For preparative purposes, larger amounts of plasmid DNA (2 μg) and, correspondent, more units of enzyme (10 Units) were used. For simultaneous restriction with *SstI* and *EcoRI*, the reaction mixture contained the buffer where both enzymes showed 100 % activity (REact2). The plasmid DNA restriction occurred by the incubation with the two restriction enzymes (2 h at 37°C). After restriction, DNA fragments were separated by gel electrophoresis (Subsection 10.1.4). When necessary, reaction was stopped by

Table 10.1: Restriction enzymes used in this work

Enzyme	Restriction site	Buffer	Supplier
BamHI	G/GATCC	REact3	Invitrogen
EcoRI	G/AATTC	REact2	Invitrogen
SstI (*SacI*)	GAGCT/C	REact2	Invitrogen
ApaLI (*Alw44I*)	G/TGCAC	NEbuffer	NEB

Table 10.2: Plasmid DNA restriction reaction components, respective amounts and final concentrations.

Reagent	Amount	Final concentration
DNA	100 - 200 ng	5 - 10 $\frac{ng}{\mu L}$
10x buffer	2 μL	1x
restriction endonuclease	1 - 2 Units	0.05 - 0.1 $\frac{U}{\mu L}$
100x BSA (when necessary)	0.2 μL	1x
LAL-H_2O	add to 20 μL	

heat inactivation according to enzymes' supplier instructions.

For the preparation of linear DNA for nucleofection, 20 μg of DNA were incubated in a reaction mixture with a total volume of 200 μL containing 100 units of the restriction enzyme *ApaLI*. The restriction occurred by incubation for 4 h at 37°C. Afterwards, a 100 ng sample was analysed by gel electrophoresis (Subsection 10.1.4). After confirmation of DNA linearization, additional purification steps with Phenol-Chloroform extraction and ethanol precipitation were done in order to remove enzyme and salts from the DNA.

10.1.4 AGAROSE GEL ELECTROPHORESIS

Agarose gel electrophoresis was used to separate, identify and purify DNA fragments by size. DNA migration depends of the molecular size of DNA, concentration of agarose, conformation of DNA (superhelical circular, nicked circular or linear), applied voltage and electrophoresis buffer. In an electric field, the negatively charged DNA moves towards cathode at rates that are inversely proportional to the logarithm of its molecular weight [HELLING *et al.* 1974]. Solutions and buffers are described in Table 10.3 Depending of the DNA fragments sizes, agarose concentrations of 0.8 or 1 % (w/v) were used. TAE buffer was used to dissolve agarose and for the run. DNA samples were mixed with 5x DNA loading buffer and loaded into the gel slots. One μg of the 2-Log DNA Ladder (0.1-10.0 kb; NEB) was used as DNA marker. Gels were regularly run for 45 min at 100 V, depending of the degree of separation necessary. Detection of DNA was achieved by staining the gel for 15 min in a bath containing 0.5 $\frac{\mu g}{mL}$ of the intercalating fluorescence dye ethidium bromide. To minimize background fluorescence, the gel was incubated for 10 min in water. Afterwards, DNA was visualized under a transilluminator with 302 nm UV light.

Table 10.3: Buffers and solutions used in Gel Electrophoresis

Buffer / Solution	**Composition**
5x DNA loading buffer	30 % (v/v) Glycerol
	0.05 % (w/v) Bromophenol blue
	0.05 % (w/v) Xylene cyanole
	0.05 % (w/v) SDS
	50 mM EDTA.2H_2O.2Na
	dissolve in 10xTAE buffer, store at -20°C
Ethidium bromide staining solution	0.5 $\frac{\mu g}{mL}$
TAE buffer	40 mM Tris
	40 mM Acetic acid
	1 mM EDTA

10.1.5 PURIFICATION OF DNA

For subsequent cloning steps, the bands of interest were excised from the gel with a scalpel and purified. For this purpose, the QIAquick Gel Extraction Kit (Qiagen) was used according to the manufacturers' instructions (QIAquick® Spin Handbook 07/2002). The principle of this kit is that DNA adsorbs to the silica-membrane in the presence of high salt while impurities are efficiently washed out. The pure DNA was eluted with 30 μL LAL-water (BioWhittaker Inc.) and stored at -20°C.

10.1.6 PHENOL-CHLOROFORM EXTRACTION

Phenol-Chloroform extraction relies on phase separation of a mixture of aqueous solution and a solution containing water-saturated phenol, chloroform and isoamyl alcohol resulting in an upper aqueous phase and a lower organic phase. Phenol gives a nebulous interface, which is sharpened by the presence of chloroform, and the isoamyl alcohol reduces foaming. Nucleic acids remain in the aqueous phase and proteins lie at the interface and in the organic phase.

One volume (equal to the volume of the DNA sample) of phenol was added to the DNA solution and the two phases were mixed for 30 s by vortexing. After spinning down (5 min at 400x *g*), the aqueous phase was transfered into a new tube. One volume of phenol-chloroform-isoamyl alcohol (25:24:1) was added and a new mixing and centrifugation step followed. The aqueous phase was pipetted into a new tube and the phenol taken up by the aqueous phase was removed by adding one volume of chloroform and subsequent mixing and centrifugation. In a last purification step, DNA was recovered from the aqueous phase by ethanol precipitation.

10.1.7 ETHANOL PRECIPITATION

This technique was used to desalt nucleic acid preparations in aqueous solutions. The basic principle is that salt and ethanol are added to the DNA solution making it less hydrophilic and thus causing the nucleic acid to precipitate out of the solution.

DNA solution was diluted with LAL-water (BioWhittaker Inc.) to a final volume of 400 μL. The precipitation took place by addition of sodium acetate to a final concentration of 0.3 M (3 M stock solution, pH 5.2) and one volume of cold 98 % (v/v) ethanol. After mixing the components, the precipitation was facilitated by incubation for 15 min at -20°C. Precipitated DNA was separated from the rest of the solution by centrifugation (15 min at 13000 rpm, 4°C). The pellet was rinsed from residual salts with cold 70 % (v/v) ethanol and, after another centrifugation step, the ethanol was discarded. The pellet was allowed to dry under a clean-bench before being eluted in 25 μL TE buffer and DNA concentration and purity were analyzed (Subsection 10.1.11).

10.1.8 LIGATION OF DNA MOLECULES

Ligation of plasmid DNA fragments was performed with T4 DNA ligase and the 10x ligation buffer supplied by NEB. DNA concentration of vector and insert were estimated taking into account the amount of plasmid DNA used for the restriction reaction, the volume used for analytical purposes and admitting a 70 % recovery after purification steps. The following relation (Eq. 10.1) was used for calculating the amount of vector and insert for the ligation reaction.

$$m_I = 3 \cdot m_V \cdot \left(\frac{bp_I}{bp_V}\right) \tag{10.1}$$

m_I	mass of insert fragment	[ng]
m_V	mass of vector fragment	[ng]
bp_I	size of insert fragment	[bp]
bp_V	size of vector fragment	[bp]

Table 10.4 indicates the reagents and respective amounts used.

Table 10.4: Restriction enzymes used in this work

Reagent	Amount	Final concentration
Vector DNA	50 ng (cohesive)	$2.5\,\frac{ng}{\mu L}$
Insert DNA	see Eq. 10.1	
10x ligase buffer	$2\,\mu L$	1x
T4 DNA ligase	$1\,\mu L$	$20\,\frac{U}{\mu L}$
LAL-H_2O	add to $20\,\mu L$	

The reaction mixture was incubated for 15 min at room temperature followed by a heat inactivation step of 10 min at 65°C. The ligation reaction was kept on ice until transformation (Subsection 10.1.9).

10.1.9 TRANSFORMATION

In molecular biology, transformation is the genetic alteration of a cell resulting from the uptake, genomic incorporation and expression of foreign genetic material (DNA).

For the transformation of NEB 5-alpha F I^q cells (NEB), the high efficiency transfection protocol was followed. Competent cells were thawed on ice for 10 min and gently mixed with $5\,\mu L$ of an ice cold ligation. After incubation on ice for 30 min, the cells were heat-shocked for 30 s at 42°C in a thermoblock. After which, the cells were incubated on ice for 5 min. Finally, $950\,\mu L$ of SOC medium (NEB) at room temperature were added and cells were incubated for 1 h at 37°C and 180 rpm. Appropriate volumes of transformed cells were plated onto pre-warmed LB plates with 50 $\frac{\mu g}{mL}$ ampicillin. The plates were then incubated overnight at 37°C.

10.1.10 PLASMID ISOLATION

The QIAprep Spin Miniprep Kit (Qiagen) was used for the small-scale isolation of plasmid DNA from bacteria, like for further applications in DNA restriction reactions or sequencing. The EndoFree Plasmid Maxi Kit (Qiagen) was used when higher plasmid amounts were required (*eg.* Transfection). Plasmid isolation was performed according to manufacturers' instructions for both kits (QIAprep Handbook 11/2005 and EndoFree Plasmid Handbook, 11/2005, respectively). The plasmid isolation principle of the QIAprep Spin Miniprep Kit is based on the alkaline lysis of the bacterial cells followed by adsorption of the plasmid DNA onto a silica-gel membrane in the presence of high salt concentration. The DNA is washed and eluted with a low concentration salt buffer. In the EndoFree Plasmid Maxi Kit, after alkaline lysis, plasmid DNA is purified by anion-exchange chromatography in the presence of high concentration salts. An additional desalting step is achieved by isopropanol precipitation.

For plasmid mini-preparation 2 mL of overnight culture were used and the purified DNA was eluted in 50 μL EB buffer (supplied with kit; Qiagen). For plasmid maxi-preparation, 250 mL of main culture were used for low copy plasmids. Purified DNA was diluted in 200 μL endotoxin-free TE buffer (supplied with kit; Qiagen).

10.1.11 DETERMINATION OF DNA CONCENTRATION AND PURITY

For quantification of DNA, UV absorption was measured at 260 and 280 nm. Nucleic acids have an adsorption maximum at 260 nm and, therefore, measurements at this wavelength permit the calculation of the concentration of DNA in the sample. One OD_{260} unit corresponds to 50 μg/mL of dsDNA. The ratio of absorbance at 260 and 280 nm was used as a test for contamination of DNA with proteins. Pure DNA preparations have $\frac{OD_{260}}{OD_{280}}$ values of 1.8. Protein contamination decreases these value, as proteins have an adsorption maximum around 280 nm [SAMBROOK AND RUSSELL, 2001].

The nucleic acid concentration was determined with the NanoDrop ND-1000 Spectrophotometer (Thermo Fisher Scientific Inc.). A 1.5 μL sample was pipetted onto the end of a fiber optic cable. A second fiber optic cable was then brought into contact with the liquid sample causing the liquid to bridge the gap between the fiber optic ends. A pulsed xenon flash lamp provides the light source and a spectrometer utilizing a linear CCD array was used to analyze the light after passing through the sample. The instrument is controlled by PC based software, and the data is logged in an archive file on the PC.

10.2 PRIMER DESIGN

A primer is a oligonucleotide, which is necessary for the confirmation of certain DNA sequences with the PCR method. In the design of these small DNA sequences, several parameters have to be taken into account to assure specificity and efficiency of the PCR technique. Among the most critical are: primer length, *GC* content, melting temperature (T_M), formation of dimmers and GC clamp. Primers for sequencing purposes were chosen to have a length of 20 to 22 nucleotides and a *GC* content of about 50 %. The melting temperature of a primer is defined as the temperature at which 50 % of the double strand DNA (dsDNA) has denatured. The annealing temperature (T_A) is the temperature at which the primer anneals to a single-stranded DNA template and is considered as 5°C lower than the T_M [SAMBROOK AND RUSSELL, 2001]. The melting temperature was between 60 and 80°C and did not differ more than 5°C between primer pairs. Primers were designed with no intra-primer homology beyond 3 base pairs. It was also considered the inclusion of more than one *G* or *C* residue at 3' end of primers to ensure correct binding.

All primers were designed with the program Clone Manager Suite 7 (Sci-Ed Software). Table 10.5 give an overview of the resulting primers used for sequencing of hCAII gene in the pIRES2-ZsGreen1 vector. These primers were purchase from Invitrogen, dissolved in LAL-H_2O to a final concentration of 10 $\frac{\text{pmol}}{\mu\text{L}}$ and stored at -20°C. The orientation of a primer is indicated by "+" for a sense primer and "-" for an antisense primer. The location of the primers hybridization region is also indicated. The melting temperatures (T_M) were calculated with the program Clone Manager Suite 7 (Sci-Ed Software).

Table 10.5: Primers for sequencing hCAII gene in phCAII-ZsGreen1

Primer	+/-	Location in phCAII-ZsGreen1	Sequence (5' → 3')	T_M [°C]
PR07	+	569	GGTTTAGTGAACCGTCAGATCC	60
PR08	+	1251	AAAGGGCAAGAGTGCTGACTTC	62

10.2.1 MYCOPLASMA TEST WITH PCR

Mycoplasmas are considered as widespread contaminants found in cell cultures, being the smallest and simplest prokaryotes that reside in endosome of mammalian cells. They can affect the cultured cells by intervening/altering their metabolism, provoking chromosomal abberations and changes in cell growth. In the present work a PCR method described by JUNG *et al.* (2003) was used for the mycoplasma detection. With this method, a spacer region between 16S and 23S ribosomal RNA (rRNA) genes is amplified through the used of primers that hybridize with those RNA regions highly preserved in prokaryotes. The size and sequence of the amplified fragment, which is different between mycoplasma species, was identified by gel electrophoresis. The length of the fragment should range from 369 to 681 bp [JUNG *et al.*, 2003]. The primers' sequences are shown in Table 10.6.

Table 10.6: Primers for PCR mycoplasma test [JUNG *et al.*, 2003]

Primer	+/-	Location (5' → 3')	Sequence [°C]
MYCF1	+	16S rRNA	ACACCATGGGAGCTGGTAAT
MYCR1	-	23S rRNA	CTTCATCGACTTTCAGACCCAAGGCAT

PCR is a method that permits the *in vitro* amplification of specific DNA fragments in a short time. It involves the following three temperature-dependent steps: denaturation of the double strand (ds) template, annealing of the primers to the single strand (ss) target sequences and replication of the DNA from the annealed primer by a polymerase. With this protocol, DNA denaturation occurs at 95°C, while annealing temperature should be between 50 and 60°C, depending of the primer used. Finally, the extension step is carried out at a temperature of optimal enzyme activity for the respective polymerase [SAMBROOK AND RUSSELL, 2001]. In this work the thermostable enzyme Taq-DNA-polymerase was used. It was originally isolated from the bacteria *Thermus aquaticus* [GELFAND *et al.*, 1989]. The enzyme exhibits optimal catalytic activity at 72°C (PeqGOLD Taq-DNA-Polymerase manual v01.07, peqLab Biotechnologie GmbH).

Cells that had been cultured for at least 3 days were collected and the cell suspension was centrifuged down at 200x *g* for 5 min. Supernatants were used as PCR template. Positive and negative controls were included in the PCR experiment to validate the method and to ensure absence of contamination, respectively. As positive control, cell-free medium from a mycoplasma infected cell culture was used. LAL-H_2O water was used as negative control. The standard PCR mixture and the different cycling parameters use for the mycoplasma detection test are listed in the Table 10.7 and Table 10.8, respectively.

Table 10.7: Standard mycoplasma test PCR mixture

Reagent	Amount	Final concentration
Polymerase buffer S	2 μL	1x
dNTP-Mix	0.5 μL	1 mM
Primer MYCF2	0.5 μL	0.25 μM
Primer MYCR2	0.5 μL	0.25 μM
Taq polymerase	0.5 μL	0.125 $\frac{U}{\mu L}$
Template	1 μL	
LAL-H_2O	add to 20 μL	

Table 10.8: PCR cycle for mycoplasma test

Step	Number of cycles	Temperature	Time
Initial denaturation	1x	95°C	3 min
Denaturation		95°C	30 sec
Annealing	35x	55°C	40 sec
Extension		72°C	40 sec
Final extension	1x	72°C	7 min
Storage	1x	4°C	∞

For analysis of the PCR-products, 4 μL of each reaction were used in agarose gel electrophoresis (see Subsection 10.1.4).

10.2.2 SEQUENCING

The sequencing of hCAII was performed at the Center for Biotechnology (CeBiTec, Bielefeld, Germany).

DNA was obtained from a Maxipreparation (Subsection 10.1.10) and delivered at a concentration of 250 $\frac{ng}{\mu L}$ in LAL-H_2O. Oligonucleotides were designed and prepared as described in Section 10.2. Nucleic acid and protein multiple sequence alignment were performed with the Clone Manager Suite 7 program (Sci-Ed Software).

10.3 NUCLEOFECTION OF MAMMALIAN CELLS

Nucleofection is an electroporation-based technique consisting of a combination of optimized solutions and electrical parameters that are supposed to deliver plasmid DNA straight to the cell nucleus and trigger rapid expression of transgenes. Nucleofection was performed with linearized DNA, to increase integration efficiency by avoiding random opening of the circular plasmid vector. The later can break the promoter, polyA signal or other important element for the correct expression of the gene of interest in mammalian cells lines. The preparation of linear DNA is described on Sections 10.1.3, 10.1.6 and 10.1.7. Nucleofection of CHO cells was performed with the Nucleofector™ II according to the protocol provided by the manufacturer (Amaxa Biosystems). In short, $5x10^6$ cells were harvested, gently mixed in 100 μL of Cell Line Nucleofector® Solution V and 4 μg of linear DNA and transferred into a 0.1 cm gap cuvette. Immediately after, the cells were pulsed with the program U-024 of the Nucleofector ™ II(Amaxa Biosystems), then pippeted into a 15 mL falcon tube containing 10 mL pre-warmed HyQ medium (Section 10.5.2) and spun down at 200x *g* for 10 min at RT. The pellet was ressuspended in 5 mL pre-warmed HyQ medium and transferred into a T-25 flask. After 24 h incubation at 37°C and 5 % CO_2, the cells were assayed for both viability and ZsGreen1 expression (transfection efficiency). After 48 h, the clone mixture was submitted to a limited dilution for single clones isolation (see Subsection 10.6.2).

10.4 CELLS

10.4.1 BACTERIA

In the present work, the high efficiency chemically competent 5-alpha F *I*q *Escherichia coli* from NEB were used.

10.4.2 MAMMALIAN CELLS

A Chinese Hamster Ovary (CHO) cell line producing a recombinant protein was kindly provided by Roche Diagnostics GmbH (Penzberg, Germany). The cell line was created by transfecting the dihydrofolate reductase (DHFR) negative CHO cell line with a recombinant vector containing genes for both a DHFR and a recombinant protein. The transfected genes were amplified by the DHFR inhibitor and methotrexate (MTX). A high producer was selected and cloned. The final cell line also contains resistance against hygromycin. This cell line was used in this work as starting point for the generation of a cell line with a better robustness against unphysiological pCO_2 levels.

10.5 MEDIA AND SUPPLEMENTS

10.5.1 MEDIA FOR CULTIVATION OF *E. coli* STRAINS

All media were sterilized by autoclaving at 121 °C for 30 min. If required, Kanamycin or Ampicillin were added to a final concentration of 50 $\frac{\mu g}{mL}$ or 100 $\frac{\mu g}{mL}$, respectively, when the media had cooled down to about 50°C. All media were stored at 4°C. The LB (Luria-Bertani) and LB agar media were used for the cultivation of *E. coli* in suspension and in static culture, respectively.

10.5.2 MEDIA FOR CULTIVATION OF MAMMALIAN CELLS

Pre-culture medium containing methotrexate was used for inoculum preparation and the main cultivations were performed in production medium. The detailed composition

Table 10.9: Composition of media for cultivation of *E. coli* strains

Medium	Composition
LB medium	10 $\frac{g}{L}$ Peptone 10 $\frac{g}{L}$ Yeast extract 2.5 $\frac{g}{L}$ NaCl adjust pH to 7.0 with 1 M NaOH
LB Agar medium	LB medium 15 $\frac{g}{L}$ Agar

of these media is confidential. The different components were dissolved in MilliQ-H_2O according to instructions von Roche Diagnostics. The pH value of the medium was titrated to pH 7.2 with 25 % (v/v) HCl. After sterile filtration (0.2 μm) and supplementing, the medium was stored at 4°C. A sterile test was performed for 48 h at 37°C and the medium was used no longer than 4 weeks after production. The commercial available HyQ® SFM4CHO™-Utility medium (Hyclone) was used as outgrowth medium after nucleofection.

10.5.3 MONOD KINETIC

Following are the formulas for calculating the specific rates. These are necessary for the evaluation of the bioreactor cultivations and are derived from the mass balances for batch cultivations [MUTZALL, 1993]. Since for this type of cultivation strategy there is no flow in or out from the bioreactor the volume is considered constant (Eq. 10.2. The cell concnetration (X) of a growing culture is changing with time (t) according to the first Eq. 10.2, which after integration and taking the logarithm originates the equation for calculating the specific growth rate (μ). The medium culture substrates are consumed as result from cell growth. They are used by cells for the maintenance of metabolic processes and for product formation. The specific substrate consumption (q_S) and product formation rates (q_P) are calculated by Eqs 10.4 and 10.5, respectively, with the mean cell concentration (X_m) calculated according to Eq. 10.6.

$$\frac{dV_L}{dt} = 0 \tag{10.2}$$

$$\frac{dX}{dt} = \mu.X \Longrightarrow \mu = \frac{\ln(X_{i+1} - X_i)}{t_{i+1} - t_i} \tag{10.3}$$

$$\frac{dS}{dt} = q_S.X \Longrightarrow q_S = \frac{1}{X_m}.\frac{S_{i+1} - S_i}{t_{i+1} - t_i} \tag{10.4}$$

$$\frac{dP}{dt} = q_P.X \Longrightarrow q_P = \frac{1}{X_m}.\frac{P_{i+1} - P_i}{t_{i+1} - t_i} \tag{10.5}$$

$$\text{with } X_m = \frac{X_{i+1} + X_i}{2} \tag{10.6}$$

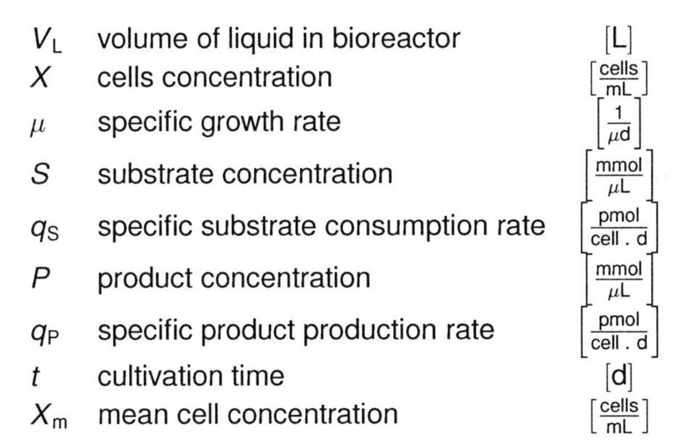

V_L	volume of liquid in bioreactor	$[L]$
X	cells concentration	$\left[\frac{cells}{mL}\right]$
μ	specific growth rate	$\left[\frac{1}{\mu d}\right]$
S	substrate concentration	$\left[\frac{mmol}{\mu L}\right]$
q_S	specific substrate consumption rate	$\left[\frac{pmol}{cell\,.\,d}\right]$
P	product concentration	$\left[\frac{mmol}{\mu L}\right]$
q_P	specific product production rate	$\left[\frac{pmol}{cell\,.\,d}\right]$
t	cultivation time	$[d]$
X_m	mean cell concentration	$\left[\frac{cells}{mL}\right]$

10.5.4 ESTABLISHING OF WORKING CELL BANK (WCB)

A working cell bank (WCB) for the original cell line (*Control*) was setup in order to make sure that the cell line remains in prime conditions and to ensure sufficient cell resources for all experiments done in this project. Therefore, cell density and viability were measured for an exponential growing culture (Subsection 10.8.2). Afterwards, cells were harvested and cell pellet was resuspended in a defined volume of chilled pre-culture medium with 5 % DMSO (dimethyl sulfoxide) in such a way, that the final cell concentration was $1x10^7 \frac{cells}{mL}$. Four milliliters of this cell suspension were aliquoted in cold 4.5 mL nominal volume cryovials (Nunc®) and cell suspension was freezed with the IceCube 1800 (SY-LAB GmbH) according to manufacturer's instructions (Instructions Manual for Ice Cube 1800, V4.01 from 08.01.2001, SY-LAB). The cryovials were transferred to the gas phase of the liquid nitrogen for long time preservation.

The WCB of the hCAII-clones or *Control HyQ*, were done either with HyQ freezing medium or with pre-culture freezing medium, both containing 5 % DMSO. Therefore, $1x10^7 \frac{cells}{vial}$ were frozen overnight at -80°C in the 5100 Cryo 1°C Freezing Container (Nalgene®) which induces a cooling rate of -1 $\frac{°C}{min}$. Afterwards, cryovials were transferred to the the gas phase of the liquid nitrogen.

10.6 CULTIVATION

10.6.1 CULTIVATION IN DISPOSABLE BIOREACTORS

Back-up or inoculum cultures and parallel test batches were carried out in disposable cultivation systems, such as bioreactor tubes and shaking flasks. In this project 50 mL conical polypropylene bioreactor tubes (TPP) and different nominal volume (125, 250 and 500 mL) polycarbonate erlenmeyer flasks (Corning) were used. Both were equipped with polyethylene plug seal caps with integrated 0.22 μm hydrophobic membrane to maintain sterility and facilitate gas exchange. For bioreactor tubes the working volume ranged between 10 and 20 mL and for shaking flasks between 40 and 60 % of the nominal volume. The cultivations were performed in an incubator (Mytron) at 37°C, 5 % CO_2 and 80 % humidity. Agitation was achieved by orbital shaking at 125 rpm (Innova 2300, New Brunswick Scientific) or 185 rpm (ES-X, Kuhner), and 0° inclination angle for bioreactor tubes.

Back-up cultures

One WCB cryovial of the control cell line (*Control*) was quickly thaw in a 37°C water bath and the cell suspension was mixed with 10 mL pre-culture medium (RT). After centrifugation (200x *g*, 10 min) to remove the DMSO, the pellet was resuspended in 10 mL RT pre-culture medium. After cell density and viability measurement (Subsection 10.8.2), a 125 mL shaking flask was inoculated at 5x10^5 $\frac{cells}{mL}$.

Clones or *Control HyQ* thawing procedure was similar to the one described for the *Control* cell line, except that after pellet resuspension in 15 mL fresh pre-culture medium, cells were cultured in 50 mL bioreactor tubes at 185 rpm (ES-X, Kuhner), and 0° inclination angle. After 2 days, cells were transferred to a 125 mL shaking flask.

Sub-culturing was performed every 3 days from an initial cell density of 4x10^5 $\frac{cells}{mL}$ for *Control* cell line or 6x10^5 $\frac{cells}{mL}$ for the hCAII over-expressing clones and *Control HyQ*. Back-up cultures were run for maximal 1 month.

Parallel batches

Parallel cultivations were used for the screening of clonal cell lines expressing hCAII gene. Cells were seeded at 6x10^5 $\frac{cells}{mL}$ and cultivated in pre-culture medium in 250 mL shaking flasks with 90 mL working volume at 185 rpm. After a 2 day adaptation phase, the main culture was inoculated in production medium. Samples of 2 mL were taken daily for cell density analysis (Subsection 10.8.2), 1 mL was centrifuged (200x *g*) for 10 min at RT. The ammonium concentration in the cell free-sample was measured (Subsection 10.8.5) and the remaining supernatant was aliquoted and stored at -20°C for glucose, lactate (Subsection 10.8.3), and product analysis (Subsection 10.8.7). The cultivations were aborted when viability dropped below 40 %. The cells were sterilely harvested and 40 mL of the cell-free supernatant were stored at -20°C for glycan analysis (Subsection 10.8.8).

10.6.2 CLONE SCREENING

For the generation of clonal cell lines, the clone pool was diluted in such a way that single cells were obtained by cultivation in 96-well plates under selective pressure. This process is called limited dilution. Forty eight hours after nucleofection, cell density and viability were checked. The clone pool was spun down and the pellet resuspended to a concentration of $2x10^3$ $\frac{\text{viable cells}}{\text{mL}}$ in selective medium (HyQ medium with 5 % FBS and 400 $\frac{\mu g}{mL}$ G418). Further dilutions were performed in order to inoculate 96-well plates (flat-bottom, Nunclon^{TM}, Nunc) with 100, 20, 10 and 5 viable cells in each well with 150 μL selective medium. The plates were incubated at 37°C and 5 % CO_2 for formation of colonies. After 7 days, 50 μL of selective medium was added the wells. Between 8 and 12 days selection, plates were checked regularly and wells containing single colonies were marked. Seventeen days after the selection procedure had started, single colonies were trypsinized by first aspiring medium and after a washing step with PBS, the cells were incubated for 5 min at 37°C with 50 μL 1x trypsin (Invitrogen). Afterwards, tripsinized cells were transferred to 24-well plates for further cultivation without selective pressure. When the colony size permitted it, the clones were further cultivated in 35 mm dishes. Once 90 % confluency was reached, part of the cells were harvested and prepared for hCAII-expression analysis (Section 10.10.1), the other part was further expanded in t-25 flasks, t-75 flasks, bioreactor tubes and, finally in 125 mL erlenmeyer flasks. In the middle of the exponential growth phase cells were split, half of which were cryopreserved in HyQ medium with 5 % DMSO (Subsection 10.5.4). The other half was centrifuged down, resuspended in pre-culture medium and further cultivated until middle exponential phase. Cryopreservations were performed in pre-culture medium with 5 % DMSO. The control cell line was cultivated throughout the screening process in HyQ medium and cryopreserved accordingly.

10.6.3 BIOREACTOR CULTIVATIONS

Batch cultivations were performed in the benchtop multi-bioreactor system Biostat® B-DCU (Sartorius AG) with 1.5 L working volume in production medium. The experimental setup is described in the following paragraphs.

The on-line monitoring of the process parameters was performed with in-place sensors, such as: pH electrode, pO_2 electrode, pCO_2 electrode and temperature electrode. The process parameters were controlled through the DCU local controller (Sartorius AG). The pH value was automatically adjusted to 7.20±0.01 by addition of 1 M NaOH or 1 M HCl. Bioreactor gassing was performed with a sterile gas mix (0.2 μ inlet sterile filter) by a ring sparger, immersed in the culture liquid under the impeller. Exhaust gas was cooled down by an exhaust gas cooler and left the bioreactor through a outlet sterile filter. The bioreactor was equipped with two 4-bladed rushton impellers. For pO_2 controlling, a cascade with gas flow rate (GAFR) and agitation (STIR) was used in the ratio mode with total flow rate of 60 ccm. A mixed gas containing air and N_2 was was used to maintain the dissolved oxygen concentration at 30.0±0.0 % air saturation. The agitation is a parameter incorporated in the pO_2 controlling and therefore, when the air flow rate reached the maximum value, this second cascade parameter, was activated and controlled according to cells oxygen demands. The initial agitation value was 80 rpm. An additional mix of CO_2 and N_2 was done to control the pCO_2 at the desired concentrations (constant 5 % CO_2 or continuous CO_2 profile from 5 % to 25 % CO_2). The bioreactor temperature was maintained at 37°C and controlled through the jacket water temperature. To exclude osmolalities differences between the CO_2-controlled cultivation modus, this parameter was manually compensated with addition of 1 M NaCl (equivalent to 1843 $\frac{mOsmol}{Kg}$) through a calibrated peristaltic pump. To avoid formation of foam, 1 % anti-foam solution was added when necessary. The reactors with the probes in place and filled with 1.5 L PBS was submitted to a pressure test. Afterwards, steam-sterilized at 121°C for 50 min.

The inoculum for the main-culture was prepared in the bioreactor at 5 % CO_2 and in production medium until it reached the middle exponential growth phase. The main cultures were inoculated at $6x10^5$ $\frac{cells}{mL}$. For sampling, the sterile coupling system with Luer-Lock (DASGIP GmbH) was used. Samples were taken every 24 h for cell growth, productivity, cellular metabolism, cell physiology and glycosylation. For sampling, syringes with the Luer-Lock system were used together with the Luer-Lock system built up in the bioreactor. To take a representative sample of the culture, a pre-flow sample (*ca.* 5 mL sample) was taken, after which, enough volume of cell suspension was removed and directly measured in the blood gas analyzer (*ca.* 100 μL) for off-line analytic (pH, pCO_2; Subsection 10.8.9) and for cell density determination (Subsection 10.8.2). Part of the remaining cell suspension was used for generation of samples for cell cycle analysis ($2x10^6$ cells, Subsection 10.9.4) and hCAII expression level ($1x10^6$ cells; Subsection 10.10.1) during the cultivation, the other part was centrifuged at 200x *g* for 10 min at RT. The ammonium concentration in the cell free-sample was measured (Subsection 10.8.5) and the rest was aliquoted and stored at -20°C for glucose, lactate (subsection 10.8.3), amino acids (Subsection 10.8.4) and product analysis (Subsection 10.8.7). When necessary, $1x10^8$ cells were harvested for proteome analysis (Section 10.11). For pH_i measurement, *ca.* 2.5 mL cell suspension was taken (Subsection 10.9.2). Finally, the sampling system was sterilized with 70 % ethanol.

The cultivations were performed until viability decreased below 40 %. At this time, the cells were sterilely harvested from the bioreactor and 40 mL supernatant were frozen for glycan analysis (Subsection 10.8.8). The pCO_2, pO_2 and pH probes were removed and the bioreactor was filled with 0.2 M NaOH. Inactivation was done overnight (45°C, 100 rpm). Finally the reactor was cleaned with MilliQ-H_2O.

10.7 *On-line* MONITORING

10.7.1 pCO_2 SENSOR

The real time in-line dissolved CO_2 monitoring was performed with an opto-chemical CO_2 measurement system (YSI BIOVISION 8500). This system is composed of a monitor, a fiber optic cable, a probe and a sensor capsule. For the B-DCU cultivation system, a top mount pCO_2 probe for 12 mm diameter bioreactor ports was used. For measurements, a YSI 8550 CO_2 sensor capsule was inserted into the stainless steel probe. The capsule consists of a small reservoir of bicarbonate buffer covered by a gas permeable silicone membrane. The buffer contains HPTS (Hydroxypyrene trisulfonic acid), a pH-sensitive fluorescent dye. Dissolved CO_2 from the culture broth (Equation 10.8) moves across a perforated steel layer, and a permeable polymer into the buffer altering its pH (Equation 10.7). As the pH changes, the fluorescence of the dye changes, accordingly.

$$CO_2 + H_2 \rightleftarrows H_2CO_3 \rightleftarrows H^+ + HCO_3^- \qquad (10.7)$$

$$PTS^- + H^+ \rightleftarrows HTPS \qquad (10.8)$$

An optical fiber transmits two wavelengths of light a light source through the cable and probe, then through a transparent polymer and into the dye layer. The resulting light emission from the dye is transmitted back through the optical fibers to the monitor, where the dissolved CO_2 is calculated based on a ratiometric analysis of the dye's fluorescence.

For the background values settings, the probe connected to the monitor by the fiber optic cable was placed in a glass reservoir containing MilliQ-H_2O. The probe was positioned more then 2 cm off the glass flask bottom. Background values for the cable and probe were checked and stored in the monitor 8500 as described in manual. A new capsule was assembled onto the distal end of the probe and LEDs, photo-detectors and sensor capsule were controlled to its functionality. According to the manual, the reference signal values should be approximately 200 with both LEDs off. With either LED 1 or 2 on, the reference value should be greater than 4000. Prior to bioreactor sterilization, the probe was installed into the bioreactor port.

After sterilization, the reactor temperature and the agitation rate were controlled to the setpoints (37°C and 80 rpm) and the reactor was sparged with 60 ccm gas mixture containing 5 % CO_2/95 % N_2 (for 5 % CO_2 controlled cultivations) or 25 % CO_2/75 % N_2 (for CO_2 profile cultivations). During the calibration and the cultivation process, the dissolved CO_2 measurement was performed at 1 min intervals in CO_2 units. When the output was stable for at least 30 min, two off-line measurements were made with the blood gas analyser and the mean CO_2 value was used to set the single point calibration. Re-calibration was done when CO_2 values differed more than 1 %.

10.8 ANALYTICAL METHODS

10.8.1 DETERMINATION OF BACTERIA OD

The optical density (OD) was determined at a wavelength of 600 nm on a two-channel spectrophotometer (UVIKON 922, Kontron Instruments). Each sample was measured twice and the arithmetic mean value was calculated.

10.8.2 DETERMINATION OF VIABLE CELL DENSITY AND VIABILITY

Cell densities and viability of the cultures were determined with the automated cell counting system Cedex (Cell Density Examination System, Innovatis GmbH). The measurement is based on the well-established Trypan Blue dye exclusion method for determination of living and dead cells [GUDERMANN *et al.*, 1997]. Sample handling, staining, cell counting, image acquisition and analysis were performed automatically by the Cedex. The principle of measurement is the digital image recognition. For better accuracy, two measurements each of 1 mL were performed for each sample (30 images).

The samples from disposable bioreactors experiments were diluted 1:2 with PBS before measurement.

10.8.3 DETERMINATION OF METABOLITES GLUCOSE AND LACTATE

Glucose is the main substrate for most mammalian cells. The control of glucose concentration in cultivations is very important for cells' growth. Lactic acid, usually referred to as Lactate, is the main metabolic product of mammalian cells. It can accumulate to high concentrations in the medium, which can affect the pH value and show inhibitory effects above 18 mM [KURANO *et al.*, 1990]. Glucose and lactate concentrations are important parameters to evaluate the behavior of a culture.

The automatic Glucose-Lactate Analyser YSI 2700D Select was used for the measurement of these parameters in supernatant. It uses the immobilized enzyme biosensor technology for analysis. For glucose and for lactate there is an enzyme, glucose oxidase and lactate oxidase, that correspondingly transforms the metabolites in D-glucono-d-Lacton and pyruvate with formation of H_2O_2. Hydrogen peroxide is oxidized by an amperometric oxygen electrode, resulting in a current which is correlated to glucose and lactate concentrations. The instrument calibrates automatically with glucose and lactate solutions of defined concentrations.

A repeat determination was performed, for each of which 150 μL sample were necessary. All samples were diluted in PBS buffer in such a way that measured concentrations were below the concentrations of the standard solutions.

10.8.4 DETERMINATION OF METABOLITES AMINO ACIDS

Amino acids are necessary to mammalian cell cultures both for protein synthesis and as an energy source. An appropriate observation of the substrate concentrations is required for the evaluation of a cultivation and its associated growth, consumption and production rates. The determination of amino acid concentrations were carried out by reversed-phase high-performance-liquid-chromatography (RP-HPLC) by the technical personal from AG Cell culture Technology according to the method described by BÜNTEMEYER (1988).

10.8.5 DETERMINATION OF AMMONIUM

Ammonium is a metabolic product of mammalian cells. It results predominantly from the thermic degradation from glutamine. Glutamine and other amino acids metabolism also contribute for the accumulation of ammonium in medium. Ammonium is a strong cellular poison; reported in the literature is a 8 mM of ammonia considered as inhibitory [HANSEN AND EMBORG, 1994].

A specific and sensitive determination of ammonium is achieved by its derivatisation in the alkaline environment to fluorescent derivatives. In the alkaline environment ammonium is converted to ammonia. Ammonia can react in chemical reactions like a primary amine. The reaction of ammonia with ortho Phthaldialdehyde (OPA) in presence of thioglycolic acid results in an isoindole derivative, which is fluorescent under suitable conditions. Unfortunately isoindole are unstable in the alkaline environment and hydrolisate, hence the measuring signal changes over time. The isoindole derivatives were excited at 415 nm and its emission was measured at 485 nm in a spectrofluorometer (RF-551, Shimadzu). For the measurement, 1.3 mL of RT reagent (25 mg OPA, 25 mg thioglycolate in 1 ml Methanol, pH to 10.4 with 10 ml 0.4 M Sodium borate buffer) were pippeted into the 1.5 mL fluorescence cuvette (Plastibrand®, Brand) the base line/value was adjusted to zero. Subsequently, 20 μL of cell-free sample were mixed with the reagent and the reaction process was followed. The maximum emission intensity valued was noted. In order to determine ammonium concentration in the sample, a reference value (standard solution 5.56 mM ammonium) was measured, accordingly. The concentration of the ammonium in the sample was determined with the relation 10.9.

$$c_{\mathrm{smpl}} = I_{\mathrm{smpl}} \cdot \frac{c_{\mathrm{std}}}{I_{\mathrm{std}}} \tag{10.9}$$

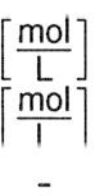

c_{smpl}	concentration of ammonium in the sample	$\left[\frac{mol}{L}\right]$
c_{std}	concentration of ammonium in the standard	$\left[\frac{mol}{l}\right]$
I_{smpl}	maximum emission intensity of the sample	-
I_{std}	maximum emission intensity of the standard	-

10.8.6 DETERMINATION OF OSMOLALITY

Osmolality is a measure for the total number of osmotically active particles in a solution and is equal to the sum of the molalities of all the solutes present in that solution. Molality is the number of particles dissolved in a mass weight of fluid $\left(\frac{\text{mmol}}{\text{Kg}}\right)$. The unit for counting is the mole which is equal to 6.02x10^{23} particles (Avogadro's Number).

The osmolality of supernatants were measured using the freezing point depression osmometer (Osmomat Auto, Gonotech GmbH). This instrument measures the change in freezing point (called, cryoscopy) that occurs in an osmotic active solution in comparison with pure water. The samples measurement was performed with 150 μL cell-free supernatant. Osmolality values were expressed in $\frac{\text{mOsmol}}{\text{Kg}}$. A two point calibration was done with MilliQ-H_2O $\left(0\,\frac{\text{mOsmol}}{\text{Kg}}\right)$ and a standard solution with predefined osmolality $\left(280\,\frac{\text{mOsmol}}{\text{Kg}} \text{ or } 320\,\frac{\text{mOsmol}}{\text{Kg}}, \text{Gonotech GmbH}\right)$.

10.8.7 DETERMINATION OF PRODUCT CONCENTRATION

The concentration of the recombinant protein was carried out by size-exclusion-high-performance-liquid-chromatography (SEC-HPLC) by the technical personal from AG Cell Culture Technology.

10.8.8 GLYCAN ANALYSIS

The glycan analysis was kindly performed by Roche Diagnostics GmbH.

10.8.9 DISSOLVED GAS ANALYSIS

The quantitative measurement of pH value and CO_2 concentrations was performed with the automatic blood gas analyzer AVL COMPACT 3 (Roche Diagnostics GmbH). A volume of 100 μL sample is necessary for the measurement. The CO_2 (%) was calculated by the Eq. 10.10

$$x_{CO2} = 100.\frac{p_{CO2}}{P_t} \tag{10.10}$$

x_{CO2}	concentration of CO_2	[%]
p_{CO2}	partial pressure of CO_2	[mmHg]
P_t	Total pressure	[mmHg]

10.9 FLOW CYTOMETRY

Flow cytometry is a technology for the quantitative analysis of multiple physical characteristics of single particles, such as relative particle size, relative granularity or internal complexity, and relative fluorescence intensity. Particles and suspended cells of 0.2 to 15 μm are suitable for analysis. A schematic drawing of the flow cytometer system composed of the three main systems (fluidics, optic and electronic) is shown in Fig. 10.4.

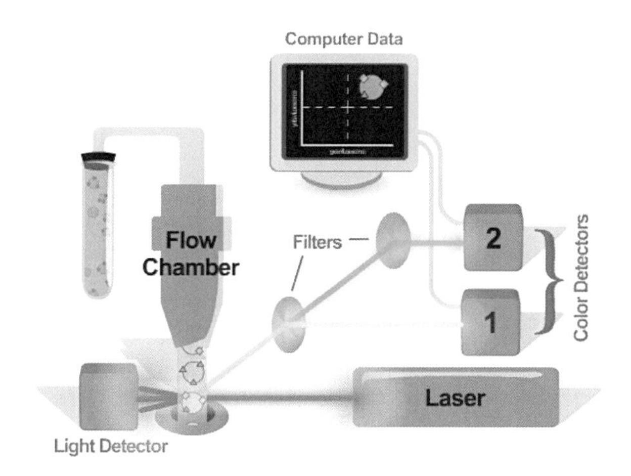

Figure 10.4: Schematic drawing of the flow cytometer system (OSU, 2009).

Cells are hydro dynamically focused in a sheath of PBS buffer (Fluidic system). As the cells pass the focused light source, they scatter light and the fluorochromes are excited to a higher energy state. This energy is released as a photon of light with specific spectral properties unique to different fluorochromes. The emitted light is then send to different detectors by using optical filters (Optical System). Scattered and emitted light from the cells is converted to electrical pulses by optical detectors (Electronic system) and this data is further processed by a computer.

This technique offers, therefore, unique insights into the gene expression and heterogeneity of the cell population in complex biological systems. In the context of this work, flow cytometry was employed for the determination of transfection efficiency (Subsection 10.9.1), intracellular pH value (Subsection 10.9.2) and cell cycle distribution of cells (Subsection 10.9.4) at different cultivation conditions. The flow cytometric analysis was carried out with a FACSCalibur™ (Becton Dickinson) using an argon laser with an excitation wavelength of 488 nm. Data acquisition and analysis was carried out using a G4 Apple Macintosh computer with the CellQuest™ Pro software. Forward light scatter (FSC) and sideward scatter (SSC) were used to examine the size and granularity of the cells.

10.9.1 DETERMINATION OF TRANSFECTION EFFICIENCY

Analysis of the *Zoanthus* sp. green fluorescent protein (ZsGreen) in CHO cells was performed by flow cytometry 24 h post-nucleofection. Briefly, $3x10^5$ cells were collected by centrifugation and resuspended in 1 mL PBS before analysis. Non-specific fluorescence was determined using non-transfected cells. Signals were acquired for the Forward Scatter (FSC-H), Sideward Scatter (SSC-H) and the FL1-H (530/30 nm) channels, using standard protocols. Samples were assayed counting 10 000 events. Through Histogram Statistics, control was gated (M1) at 99.95 % of all events and the samples transfection efficiency was determined with gate M2.

10.9.2 INTRACELLULAR PH MEASUREMENT

There is a considerable interest in measuring and monitoring the dynamics of intracellular pH value (pH_i) change to determine its implication on biological processes. In the context of this work, pH_i measurements were carried out by flow cytometry with the available pH_i fluorescent indicator, Carboxy SNARF-1 AM (Molecular Probes Inc.). This dye is an acetoxymethyl (AM) ester that enter the cells readily and is hydrolyzed by nonspecific esterases to yield the free fluorescent dye. SNARF-1 AM is very sensitive to pH_i changes within the physiological range [WIEDER *et al.*, 1993]. It is excited between 488 and 530 nm and the fluorescence emission is monitored at two wavelengths, typically about 580 nm (FL2-H) and 640 nm (FL3-H). This is particularly interesting as the fluorescence emission wavelength ratio can be calculated foe a more accurate determination of the pH_i. With the use of this ratiometric technique, a number of fluorescence measurement artifacts are eliminated, such as photobleaching, cell thickness and leakage and nonuniform loading of the indicator (SNARF®pH indicators, Product information, 2003, Molecular Probes Inc.). The calibration of the fluorescence response of the dye for each experimental system is necessary. There are different calibration protocols to equilibrate the intracellular pH value with the controlled extracellular medium. In this project, the Pseudo-Null calibration method was used. It is based on the Null Point method for calibrating spectrofluorometric pH_i measurements, as proposed by EISNER *et al.* (1989), which was in turn developed from an earlier work by SZATKOWSKI AND THOMAS (1986). These later scientists derive the expression for steady-state pH_i from the combined Henderson-Hasselbalch relations for a weak acid and a weak base. The method is based on the assumption that (i) only the non-ionized forms of the weak acid or base can cross the cell membrane, (ii) the dissociation constants, pK_a and pK_b, are the same inside and outside of the cell, and (iii) mechanisms that regulate pH_i do not influence the critical changes in pH_i (or fluorescence) observed. A simplification can be considered if the calibration solutions are prepared in such a way that a concentration of weak base and a concentration of weak acid produce an equal but opposite change in pH_i (*i.e.*, $pH_a = pH_b$). The derived relationship for pH_i determination is presented in Eq. 10.11.

$$pH_i = pH_i - 0.5 \log\left(\frac{[A]}{[B]}\right) \tag{10.11}$$

pH_i	intracellular pH	-
pH_e	extracellular pH	-
[A]	concentration of weak acid	[mM]
[B]	concentration of weak base	[mM]

Where the pH_i is dependent on the extracellular pH value(pH_e) and on the ratio of concentration of weak base (*B*) and weak acid (*A*) present in the solution. EISNER *et al.* (1989), developed this theory further by determining the mixture of weak acid and base that produces no change in the signal obtained from cells loaded with a fluorescent pH value indicator. The ratio that gives no change in pH_i could be extrapolated between those that lead to the smallest increase in pH_i ("Null Point"). This null point method is particularly useful when the indicator response is not a simple function of the pH value and is presumably independent of mechanism that regulate the pH_i, because the pH_i does not change at null point. In turn, CHOW *et al.* (1996), investigated the cells exposure to a given combination of acid and base. If the molar concentration of the acid-base ratio is sufficient, then no further addition of acid to base in the same ratio causes a change in the pH_i, so that value reflected a new null value (designated Pseudo-Null) that satisfies Equation 10.11. Therefore, a calibration curve can be obtained from the plot of pseudo-null pH value vs. fluorescence ratio by exposing cells to a series of acid-to-base mixtures at sufficient molar concentration. Those mixtures were done with butyric acid (BA) and trimethylamine (TMA) (Table 10.10). The choice of these weak acid and base are well elucidated by SZATKOWSKI AND THOMAS (1986). The preparation of calibration solutions was done in HEPES buffer containing FBS (named HDFBS buffer; Table 10.10). The addition of FBS aids maintenance of the cells physiological conditions.

Table 10.10: List of buffers and solutions used in the in the Pseudo-Null Point pH_i determination method. All solutions except, Carboxy SNARF-1 AM, were sterile filtrated. All solutions were stored at 4°C.

Buffer / Solution	Composition
Carboxy SNARF-1 AM	dissolve in DMSO at 2 μM
	store at 4°C
HEPES buffer	10 mM HEPES
	133.5 mM NaCl
	4 mM KCl
	1.2 mM $NaH_2PO_4.H_2O$
	1.2 mM $MgSO_4$
	11 mM α-D-glucose
	2 mM $CaCl_2.2H_2O$
	adjust pH to 7.4
HDFBS buffer	90 % (v/v) HEPES buffer
	10 % (v/v) FBS
Butyric acid (BA)	1 M n-butyric acid (pKa 4.82)
	adjust pH to 7.4
Trimethylamine (TMA)	1 M trimethylamine (pKb 9.8)
	adjust pH to 7.4

A 5 mL disposable syringe was pre-loaded with 22 μL of SNARF-1 AM (2 mM in DMSO). Subsequently, 2.5 mL samples were taken from a bioreactor, the syringe was closed with a Luer-stopper (Rotilabo) and cells were stained for 25 min at 37°C. Afterwards,

200 μL of stained cells were mixed with 200 μL of 37°C HDFBS buffer and measured with the FACS. Samples were analyzed in replicates. Emission fluorescences was measured at 580 nm (FL2-H) and 640 nm (FL3-H).

Calibration curves were established with the same sampled stained cells. Hence, six times concentrated standard solutions were prepared by adding different amounts of butyric acid (BA) and trimethylamine (TMA) to the HBFBS buffer, as shown in Table 10.11.

Table 10.11: Buffers used in the Pseudo-Null Point pH_i determination method. BA = butyric acid (from 1 M stock solution); TMA = trimethylamine (from 1 M stock solution)

Buffer	**6x concentration** BA/TMA (mM) in HDFBS Buffer, pH 7.4	**Pseudo Null pH**
S1	6/96	8.0
S2	6/24	7.7
S3	6/6	7.4
S4	24/6	7.1
S5	96/6	6.8

A volume of 200 μL of each solution was pipetted into the polypropylene tubes, closed with a cap to avoid evaporation and incubated at 37°C. Calibration was achieved by mixing 200 μL of dye-loaded cells and 200 μL of aliquoted and prewarmed 6x calibration solution. Flow cytometry acquisition began after 20 sec in "HIGH RUN" modus and 10000 total events were analyzed. Prior to flow cytometric analysis, two calibration mixtures with cells were measured with the blood gas analyzer (subsection 10.8.9) to obtain the exact external pH value. The emission fluorescence ratios, $\frac{640\,nm}{580\,nm}$, were calculated with the program FCSassist™ 1.0. Further data analysis was executed with the CellQuest™ Pro software. Cells that failed to retain SNARF-1 fluorescence were gated out on the fluorescence histograms. From the histogram $\frac{FL3\text{-}H}{FL2\text{-}H}$ vs. counts analysis a histogram statistic was made and the mean fluorescence value was noted. With the information of the true external pH of the calibration samples, the induced pH_i was calculated using the Equation 10.11. The mean fluorescence ratio of each standard was plotted against calculated induced pH_i. The samples pH_i was derived from the curve.

10.9.3 RE-ALKALINIZATION EXPERIMENT

Human CAII and NHE1 isoform of the mammalian Na^+/H^+ exchanger form a complex *in vivo* and from this interaction results the regulation of the transport by NHE1 [LI *et al.*, 2002 and 2006]. These scientists tested and validate the hypothesis of cells transfected with both hCAII and NHE1 genes. To find out if this theory is true for the generated hCAII-overexpressing cell line, the regulation of the pH_i was studied by flow cytometric analysis during intracellular acidification with bicarbonate. The kinetic of the transporters were examined. In this study, this test was made at physiological conditions. Therefore medium was used instead of buffer to account for influences of other medium components. To eliminate the activity of other cellular pH_i-regulation systems, cells were pre-incubated in bicarbonate-free medium buffered with HEPES. A closed-system was built to avoid CO_2 degassing that could cause pH_e and, hence ensuing pH_i alterations. This system consisting of two disposable 5 mL Luer-Lock syringes (BBraun AG) connected by a short tubing and two Luer tubing connectors (Rotilabo®). This Luer connectors permitted the rapid and easy connection/disconnection of the 2 syringes (Fig. 10.5).

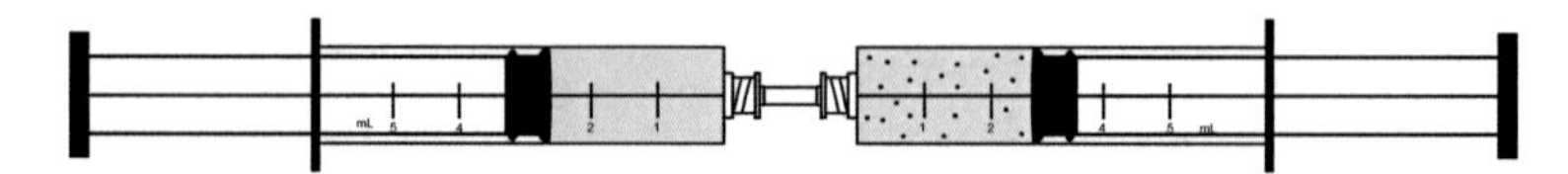

Figure 10.5: Schematic drawing of the syringe system to analyse the transporters kinetic during cells re-alkalinization.

Syringe *A* contained stained cells in bicarbonate-free production medium and syringe *B* production medium with 4.2 $\frac{g}{L}$ $NaHCO_3$ (corresponding to 25 % CO_2 at pH 7.2). Both media contained 20 mM HEPES and were adjusted to pH 7.2 at 37°C. Acid-load was achieved by pressing bicarbonate-containing medium from syringe *B* into syringe *A*. A 5 mL syringe was pre-loaded with 22 μL of SNARF-1 AM (2 mM stock solution; Table 10.10). 1x10^6 cells, from the exponential phase, were harvested and washed once with 4 mL medium without bicarbonate. After which, the cell suspension was sucked (without air bubbles) into the pre-loaded fluorescence dye syringe, which was them closed with the Luer-stopper and gently mixed. After a 25 min period of incubation at 37°C in the dark, with regularly mixing to avoid cell settling, followed dissolved gas and flow cytometric analysis. Data acquisition was done as described in Subsection 10.9.2. Three measurements of loaded cells in bicarbonate-free medium were made before acidification of cells cytoplasm. Afterwards, the syringe *B* was connected to syringe

A and the double volume of $NaHCO_3$-containing medium was added. The syringe *B* was discarded, syringe *A* was closed and after short mixing, the acid-loaded cells were analyzed by flow cytometry. Additionally, some samples were measured with the blood gas analyzer (Subsection 10.8.9) to determine the content of CO_2 in medium during the experiment.

To test if the re-alkalinization effect was due to the hCAII-overexpression, the hCAII inhibitor acetazolamide (ACTZ) was used. For the hCAII inhibited experiments, media with 100 μM acetazolamide (stock solution: 100 mM ACTZ in DMSO stored at 4°C) was used. The experiment procedure was the same as described above.

10.9.4 CELL CYCLE

The cell cycle is defined as the interval between completion of mitosis in the parental cell and completion of the next mitosis in one or both daughter cells [BASERGA, 1985]. It can be divided into four distinct phases. The G_1 phase is the period of synthesis of various enzymes that are required in *S* phase, mainly those needed for DNA replication. The *S* phase is the period of DNA synthesis, during which the genome is duplicated. The G_2 phase is the mature stage of the cell, lasting from the end of the genome duplication until the onset of the mitotic prophase. This phase is characterized by significant protein synthesis. The *M* phase is the short period of mitosis, during which extensive structural changes appear, and the end of which the division of the cytoplasm is initiated. After completing one cycle and reaching the next G_1, the cell can either proceed through another cycle or can enter the G_0 and stay there until it is stimulated. Cells that have temporarily or reversibly stopped division are said to have entered a state of quiescence [LEELAVATCHARAMAS *et al.*, 1996; KUBBIES *et al.*, 1996].

To obtain useful insights into the proliferation kinetics of CHO cells, the measurement of only a single parameter, the deoxyribonucleic acid (DNA) content, is necessary for cell cycle analysis. In permeabilized cells, the DNA measurements are achieved by staining with the fluorochrome propidium iodide (PI), after enzymatic treatment with specific nucleases to degrade the ribonucleic acid (RNA). This fluorochrome has an absorption peak at 536 nm and emission peak at 617 nm. The fluorescent molecule intercalate with double-strand DNA between the base pares (A-T, G-C). While cells in the G_1 and G_0 phases of the cell cycle have single DNA content, cells at the G_2 and *M* phases contain the double amount of DNA. In the *S* phase, were the DNA replication occurs, the DNA content is between single and double, depending of the progress of the cell inside of the synthesis phase.

Samples were taken every 24 h until viability decreased below 60 %. Approximately $1x10^6$ cells were harvested and two times washed with cold-PBS by centrifugation (200x g, 10 min, RT). Cells were then fixed and permeabilized with 1 mL ice-cold 70 % ethanol and kept at -20°C until staining. Before staining, cells were spun down followed by washing with PBS/0.1 % Saponin. Cells were then stained with 1 mL staining solution (PBS/0.1 % Saponin, 40 $\frac{\mu g}{mL}$ RNase S, 20 $\frac{\mu g}{mL}$ PI) by incubation in the dark for 45 min at room temperature. The RNA degradation reaction was stopped by incubating the stained cells on ice until flow cytometric analysis. Data was acquired with "LOW RUN" modus at a flow rate < 200 $\frac{cells}{sec}$. The G_0G_1 cell fraction was acquired on channel 200 of FL3-H (640 nm) histogram. The integral fluorescence of the cells was analysed by the computer software ModFit^{TM} *LT* (Becton Dickinson) to obtain the percentages of cells in the G_0G_1, *S*, and G_2M phase.

10.10 PROTEIN BIOCHEMICAL METHODS

10.10.1 PROTEIN EXTRACTION

Cell lysis is the first step in cell fractionation and protein purification. This step can be performed by diverse methods and/or using different buffers depending on the use of the protein extract. In the present work, cell lysis is described the sample preparation for western blotting, enzyme activity and proteomic analysis purposes.

For western blot

About $1x10^7$ cells were harvested and washed two times with cold PBS. The cell pellet was resuspended in 600 µL lysis buffer (Table 10.12) and incubated for 5 min on ice. The lysis proceed with a 5 min sonication step followed by a 30 min incubation step on ice. The lysate was subsequently centrifuged for 30 min at 16200x *g* and 4°C to remove cell debris. The protein solution was stored at -20°C.

Table 10.12: Buffers composition for cells lyse.

Buffer / Solution	Composition
Lyse buffer	50 mM Tris-HCl (pH 7.2)
	2 mM EDTA
	150 mM NaCl
	1 % (v/v) NP-40
	1 mM PMSF (add freshly from stock solution)
	0.1 % (w/v) SDS
RIPA buffer	50 mM Tris-HCl (pH 8.0)
	150 mM NaCl
	5 mM EDTA
	1 mM PMSF (add freshly from stock solution)
TE buffer	10 mM Tris-HCl (pH 8.8)
	1 mM EDTA
	2 mM PMSF (add freshly)
PMSF stock solution	100 mM phenylmethylsulphonyl fluoride in isopropanol
DNase/RNase-mix stock solution	1 $\frac{mg}{mL}$ DNase
	0.25 $\frac{mg}{mL}$ RNase
	50 mM $MgCl_2$

For hCAII activity

Ionic detergents, like SDS, tend to denture proteins by destroying their secondary, tertiary and quaternary structure, although antibody activity and some enzyme activities are retained at low concentrations (less than 0.1 % SDS). TONG *et al.* (2000) suggested that CAII activity is blocked by SDS concentrations of 0.2 %. Most proteins can tolerate levels of 1-3 % (w/v) of the non-ionic detergent Triton X-100 and still retain complete bioactivity [SAMBROOK AND RUSSELL, 2001]. Hence, the protein extraction was performed with RIPA Buffer (Table 10.12). About $1x10^7$ cells were harvested and two times washed with cold PBS. The resulting pellet was resuspended in 2 mL RIPA-buffer and solution was frozen at -80°C for 2 h. After thawing on ice, cell suspension was treated for 15 sec with an Ultrasonic finger (Sonifier 250, Branson). Cell debris was spun down (17000x *g*, 1 h, 4°C) and cytosolic proteins were stored at -20°C until carbonic anhydrase activity measurements.

For proteomic analysis

For the proteomic analysis, $1x10^8$ cells were harvested from the bioreactor at cultivation day 2 (before CO_2 profile) and 4 (two days after CO_2 profile had started) for both cell lines (Clone E11 and Control HyQ). Cells were washed with cold PBS (200x *g*, 10 min) and pellet was stored at -80°C until cell lysis. The frozen cell pellet was resuspended in 1 mL TE-buffer (Table 10.12). Subsequently, a 20 μL of a serin protease inhibitor stock solution, phenylmethylsulphonyl fluoride (PMSF), and 100 μL of a DNase/RNase-mix stock solution were added (Table 10.12). After transferring the cells in solution to a 2 mL eppi containing 1 g glass beads (ϕ 0.15 mm; BioSpec Products Inc.), the mechanical cell disruption was carried out with four homogenization cycles performed in a vortex, from 30 sec each. Between each cycle, the cells were stored on ice. Afterwards, proteins in solution were separated from the cells debris and glass beads by a centrifugation step (16200x g, 20 min, 4°C). For the separation of the proteins in solution and cells compartments, an ultracentrifugation at 106000x *g* for 1 h at 4°C (OptimaTM L-90K, Beckman Coulter Inc., USA) was performed. To determine the protein content of the supernatant part of the protein extract was removed to perform a BCA assay (subsection 10.10.2). The rest of the protein solution was stored at -20°C.

10.10.2 DETERMINATION OF PROTEIN CONCENTRATION

The determination of protein concentration was performed by the bicinchoninic acid (BCA) method using the protein quantification kit (Uptima/Interchim). It involves the reduction of Cu^{2+} to Cu^{+} by peptidic bounds of proteins. The bicinchoninic acid chelates Cu^{+} ions with very high specificity to form a water soluble purple colored complex. The reaction is measured at 562 nm, corresponding to the high optical absorbance of the final Cu^{+} complex. The protein concentration is proportional to the absorbance. It was followed the protocol described by the manufacturer with a standard curve between 20 and 2000 $\frac{\mu g}{mL}$ of Bovine Serum Albumin (BSA). Samples and standard were diluted in the corresponding lysis buffer. The absorption was measured with the Photospectrometer PowerWave^{TM} HT (BioTek Instruments). The quantification was performed with the Software KC4 (BioTek Instruments).

10.10.3 ACETONE PRECIPITATION

The acetone precipitation was carried out in order to eliminate substances that may interfere with downstream processes and to define a certain protein amount in a sample for a particular analysis. Acetone is a water-soluble solvent with small dielectric constant. When added to an aqueous solution of proteins, acetone lowers proteins solubility and, consequently, precipitation is induced.

Aliquots of 10 or 40 μg and 150 μg or 450 μg were prepared for western blotting and for proteomic analysis, respectively. Therefore, one volume of sample was mixed with 9 volumes of ice cold acetone. Subsequently, the sample was incubated overnight at -20°C. After centrifugation (16200x *g*, 30 min, 4°C) to pellet the precipitated protein, the supernatant was removed and the undesired acetone was allowed to evaporate at RT. The dried protein pellet was stored at -20°C.

10.10.4 SDS-PAGE

Proteins were separated on the basis of their molecular weight by Sodium dodecyl sulphate polyacrylamide gel electrophoresis (SDS-PAGE). Preparation of gel electrophoresis module and gel cassette was performed according to manufacturers' instructions

(Invitrogen). The protein sample pellets to be analyzed and 20 μg the positive control (carbonic anhydrase isozyme II from human erythrocytes, 1 $\frac{mg}{mL}$, Sigma Aldrich) were mixed with sample buffer (4x, NuPAGE®), reducing agent (10x, Invitrogen) and MilliQ-H_2O to a final volume of 20 μL. The denaturation of the secondary and tertiary structures and the reduction (cleaving of disulphide bons between cysteine residues) of proteins was achieved by incubating samples at 70°C for 10 min in a thermoblock. The presence of the anionic detergent SDS coats the polypeptides and masks their native charges, giving them an uniform negative charge. Finally, the samples and 5 μL of the molecular weight marker (SeeBlue® Plus2 Pre-stained standard, 4 kDa to 250 kDa; Invitrogen) were loaded into the 10 % NuPAGE® Novex Bis-Tris Gel (Invitrogen) slots. The gel was run with NuPAGE® MOPS SDS running buffer, furthermore, reducing condition were achieved with 0.25 % (v/v) antioxidant reagent in running buffer in the middle of the electrophorese camera. The gel run in the XCell™ Sure Lock system (Invitrogen) at 200 V until bromphenol reached the end of the gel.

10.10.5 WESTERN BLOT

The solutions necessary for this experiment are described on Table 10.13. Following SDS-PAGE, the separated proteins were transferred from the gel to a thin Hybond™-P PVDF-membrane. Prior to this, the membrane was previously in methanol wetted (30 sec) and equilibrated in transfer buffer. The XCell II™ Blot Module was set up according to the manufacturers instructions. Transfer buffer was added to the inside of the Blotting Module and MilliQ-H_2O to the outer part. The blotting was carried out at a constant voltage of 35 V for 1 h. Afterwards, the membrane was blocked in 20 mL blocking solution (1 h, shaking, RT), in order to prevent unspecific antibody binding in the following steps. After washing the membrane twice, the carbonic anhydrase II was detected in a three-step procedure. In the first step, the membrane was incubated (4°C) with a 1:50000 or 1:20000 specific primary antibody diluted solution (Rabbit polyclonal IgG to CAII, 10 $\frac{mg}{mL}$, Abcam) overnight. In a second incubation step (1 h RT), 1:4000 peroxidase-conjugated secondary antibody diluted solution (ECL™ Donkey anti-rabbit IgG, HRP-Linked, 0.20 $\frac{mg}{mL}$, GE Healthcare) was applied. In a final step, the membrane was exposed to the ECL™ Plus Western Blotting Reagent (Amersham), and the qualitative analysis was done by means of enhanced chemoluminescence (ECL) and autoradiography ECL-films (Hyperfilm™ ECL™ film, Amersham), according to manufacturer's instructions. Between the incubations with the different antibodies and before detection, the membrane was washed shortly with washing solution, followed by two 15 min washing steps. All washing steps were performed at RT while shaking. Film exposure times ranged from 30 sec to 5 min. Afterwards, films were developed until protein band had a good resolution, followed by a short washing step in MilliQ-H_2O and fixation for about 10 min. Later, the films were watered for about 30 min and finally dried.

Table 10.13: Solutions for western blotting.

Buffer / Solution	Composition
Transfer buffer	1x NuPAGE® Transfer buffer (20x) 0.001 % (v/v) NuPAGE® Reducing agent
10 % (v/v) Methanol	
Washing solution	0.3 % (w/v) Milk pulver 0.3 % (v/v) Tween 20 in PBS
Blocking solution	3 % (w/v) Milk pulver in washing solution
Antibody solution	1 % (w/v) Milk pulver in washing solution

10.10.6 CARBONIC ANHYDRASE ACTIVITY

Carbonic anhydrase II activity measurements were kindly done my Mr. Tu belonging to the Departments of Pharmacology and Biochemistry coordinated by Prof. Dr. Silverman (University of Florida, College of Medicine). The activity of carbonic anhydrase in lysates was done using the ^{18}O exchange method developed by SILVERMAN (1982). The ^{18}O exchange method is based on the assessment by membrane-inlet mass spectrometry of the exchange of ^{18}O between CO_2 and water at chemical equilibrium (Eqs. 10.12 and 10.13) e in lysates was done using the ^{18}O exchange method developed by [SILVERMAN, 1982].

$$HCOO^{18}O^- + EZnH_2O \rightleftarrows EZnHCOO^{18}O^- \rightleftarrows COO + EZn^{18}OH^- \qquad (10.12)$$

$$EZn^{18}OH^- + BH^+ \rightleftarrows EZnH_2^{18}O + B \underset{H_2O}{\overset{\rightleftarrows}{\longrightarrow}} EZnH_2O + H_2^{18}O + B \qquad (10.13)$$

Lysates from control cell lines (*Control* and *Control HyQ*) and from clones *G9* and *E11* were prepared by two different approaches. One of them focused on the cytoplasm protein extraction in RIPA buffer and the other on the protein extraction by mechanical disruption with glass beads in TE buffer (Subsection 10.10.1). As positive control, the commercial available carbonic anhydrase II (carbonic anhydrase isozyme II from human erythrocytes; Sigma Aldrich) was used. The lysates were divided into two and positive control was added to one of the samples in order to exclude influence of buffer composition and other proteins in the determination of CAII activity. The total enzyme concentration of a solution containing the positive control was computed using a linear regression fitting to the inhibition data for the titration curve with highly CAII bound inhibitor ethoxyzolamide (EZA). This value was used as standard for the calculation of the CAII concentration in the samples using Eq. 10.14.

$$c_{smpl} = \frac{Act_{smpl}}{Act_{std}} \cdot \frac{V_{std}}{V_{smpl}} \cdot c_{std} \tag{10.14}$$

c_{smpl}	concentration of CAII in the sample	$\left[\frac{\mu mol}{L}\right]$
c_{std}	concentration of CAII in the standard	$\left[\frac{\mu mol}{L}\right]$
Act_{smpl}	activity of CAII in the sample	-
Act_{std}	activity of CAII in the standard	-
V_{smpl}	volume of sample used for measurement	μL
V_{std}	volume of standard used for measurement	μL

10.11 PROTEOME ANALYSIS

A DIGE project was carried out to compare the clone E11 and control HyQ under different CO_2 levels at the protein level. Samples were taken from bioreactor cultivation and prepared as described in subsections 10.10.1 and 10.10.3. The first sample was taken at the time point were the cultivation conditions were comparable, i.e., before the start of the CO_2 profile (cultivation day 2), and in order to minimize the number of samples, the samples from the parallel cultivations of the same cell line were pooled after solubilization (Control HyQ: sample A, Clone E11: sample D). The second sample was taken in order to investigate the influence of a high CO_2 level in the protein expression, therefore at cultivation day 4, with CO_2 13 %, the samples B and C were taken for Control HyQ and samples E and F for Clone E11. The all project involve a total of 6 different samples (sample A to F).

10.11.1 SAMPLE PREPARATION

Classical 2D electrophoresis

The applied protein amount for the classic 2D Gel electrophoresis was dependent on the staining method chosen. In the case of Coomassie staining, 450 μg per gel were needed, whereas only 150 μg were applied for the more sensitive silver staining. The first step was the resuspension of the pellet in solubilization buffer (table 10.14). For each isoelectric focusing run, 450 μL sample were needed with the appropriate protein amount (450 μg or 150 μg). After vigorous vortexing, the samples were incubated on the shaker for 1 h at room temperature.

2D-DIGE

Each sample for the differential gel electrophoresis contained a total of 150 μg total protein from two different samples and an internal standard. The internal standard is a mixture of equal proportions of each sample used in the project. For the internal standard the CyDye^{TM}DIGE Fluor minimal dye Cy^{TM}2 (GE Healthcare) was used and the Cy^{TM}3 and Cy^{TM}5 for the samples.

For the sample preparation for 2D-DIGE, three 150 μg pellets of samples A to F were resuspended each in 150 μL of DIGE thiourea solubilization buffer "DTS" and shaken vigorously for 1 h at room temperature. Thus, per sample resulted in protein solutions of 1,μg/mL. After centrifugation (16000 x*g*, 4°C, 5 min), two 150 μL portions of each sample were mixed together and aliquoted to 100 μL. The first 100 μL were used to check pH, while the other aliquots were labeled half with 2 μL Cy3^{TM} or Cy5^{TM} dye. The third portion of each sample was used to generate the internal standard. This was aliquoted in 300 μL and 600 μL, to adjust pH and to label with 12 μL Cy2^{TM} dye, respectively. The labeling was achieved by incubation for 30 min in the dark. After stopping the reaction by addition of 2 μL or 12 μL of a 10 mM lysine solution, a further incubation for 10 min was carried out on ice in the dark.

The combination of the samples for the gels was made according to Table 10.15, 50 μL of the labeled internal standard was pippeted into a reaction vessel followed by 50 μL of differently labeled sample. A final volume of 450 μL was achieved by adding 150 μL of 2xDTS followed by 150 μL of 1xDTS with final centrifugation for 5 min at 16000 x*g* at 4°C. These samples were further used for isoelectric focusing.

Table 10.14: List of buffers and solutions used in the Proteome analysis.

Buffer / Solution	Composition
1x Solubilization buffer DTS	7 M urea 2 M thiourea 4 % (w/v) CHAPS 1 % (w/v) DTT 1 % (v/v) Pharmalyte 3-10 pH NL
2x Solubilization buffer DTS	Solubilization buffer DTS 2 % (w/v) DTT 2 % (v/v) Pharmalyte pH 3-10NL
Solubilization buffer "DTS"	7 M urea 2 M thiourea 4 % (w/v) CHAPS 30 mM Tris-HCl adjust pH to 8.5
Acrylamide solution	188 mL Rotiphorese Gel 30 113 mL of 1.5 M Tris-HCl (pH 8.8) 140 mL of MilliQ-H_2O 4.5 mL of 10 % (w/v) SDS 62 μL of tetramethylethylenediamine "TEMED" 4.5 ml of 8 % ammonium persulfate
DTT solution	10 mg/mL DTT in equilibration buffer 50 mM Tris-HCl (pH 8.8) 6 M urea 30 % (v/v) glycerol 2 % (w/v) SDS
Iodoacetamide solution	40 mg/ml iodoacetamide in equilibration buffer
Bromophenol blue	1 % (w/v) BPB in 50 mM Tris buffer
1x SDS electrophoresis buffer	25 mM Tris 192 mM glycine 0.5 % (w/v) SDS
2x SDS electrophoresis buffer	50 mM Tris 384 mM glycine 0.5 % (w/v) SDS
Agarose Sealing Solution	0.5 % (w/v) agarose in SDS electrophoresis buffer

Table 10.15: Overview of the gel layout of the DIGE project

Gel number	Cy2 dye	Cy3 dye	Cy5 dye
I	intern standard	A	B
II	intern standard	B	C
III	intern standard	C	A
IV	intern standard	D	E
V	intern standard	E	F
VI	intern standard	F	D
VII	intern standard	A	D
VIII	intern standard	B	E
IX	intern standard	C	F
X	intern standard	D	A
XI	intern standard	F	B
XII	intern standard	E	C

10.11.2 1ST DIMENSION: ISOELECTRIC FOCUSING

The isoelectric focusing (IEF) was used by both classical and DIGE approaches. The equipment and materials used were purchased from GE Healthcare. First, the entire sample was uniformly distributed between the two electrodes of a ceramic IEF vessel. Then, the IEF strip (24 cm Immobiline Dry Strip pH 3-10 NL) was applied on the protein solution with the IPG gel downwards. The gel strip was covered with approximately 1 mL of IPG cover fluid and sealed with the lid. The separation of protein mixture by charges was carried out in the IPGphor III system with the following program:

Step 1: 30 V 12 h rehydration 30 Vh
Step 2: 200 V 0.5 h step 100 Vh
Step 3: 600 V 0.5 h step 300 Vh
Step 4: 2000 V 1 h step 2000 Vh
Step 5: 8000 V 6 h step 48000 Vh

The flow of current was limited to 50 μA per strip. The protein focusing finished when reaching a total of 48000 Vh. The duration of a IEF run is, among others, dependent on the salt concentration in the samples but, on average, it lasts ca. 22 h. After completion of the focusing, the IPG strips were taken from the ceramic vessel and the excess of IPG cover fluid was removed and thereafter used for the 2nd dimension.

10.11.3 2ND DIMENSION: SDS-PAGE

Equilibration of IPG strips

Before applying the IPG strips for the SDS-PAGE, each strip was initially incubated for 15 min in 15 mL dithiothreitol solution. After washing with MilliQ-H_2O, an equilibration of each strip was carried out for 15 min in 15 mL iodoacetamide solution. 200 μL of bromophenol blue was added to each 45 mL of both solutions before use (Table 10.14).

Casting of polyacrylamide gels

Polyacrylamide gels with a size of 20x20 cm were produced for the 2nd dimension. Standard glass plates were used for casting the gels for the classical approach in contrast with special low-fluorescence glass plates for DIGE approach. The acrylamide solution was freshly prepared with 4°C cooled components according to Table 10.14 followed by 5 min degassing in an ultrasonic bath. To trigger the polymerization of the gel solution 4.5 mL of 8 % ammonium persulfate was added just before pouring. Following the casting the gels were covered with 0.1 % (w/v) SDS and the gel support was covered with a SDS soaked cloth. On the following day, the gels were polymerized and could be used.

SDS-PAGE

For the SDS-PAGE, the gels prepared on the day before were rinsed with deionized water and covered with MilliQ-H_2O. The IPG strip was briefly immersed in SDS electrophoresis buffer and with a spatula pushed between the glass plates. In the classical approach, the Precision Plus Protein^2 Standard (Cat. no. 161-0363, Bio-Rad), with eleven bands between 10 kDa and 250 kDa, used for protein size determination. For this purpose, 10 μL standard were mixed with 10 μL agarose sealing solution and pipetted onto a filter paper, which was placed next to the IPG strips on the gels and both overlaid with agarose sealing solution. The finished gel cassettes were placed in the chamber (Ettan™ DALTsix Electrophoresis Unit, GE Healthcare, USA). The lower chamber was filled with 1x SDS electrophoresis buffer and the upper chamber with 2x SDS electrophoresis buffer. A power of 5 W was chosen for the initial 45 min run, while the main run was carried out at 17 W per gel. The gel run terminated when the bromophenol blue was on the front end of the gels. After removing the gels from the cassettes, they were rinsed with deionized water. The DIGE gels were scanned (subsection 10.11.4) or stored at 4°C in wet towels. The classically produced gels were silver or coomassie stained (subsection 10.11.5).

10.11.4 GEL SCANNING

For the visualization of fluorescent proteins, the DIGE gels were scanned with the Ettan[2] DIGE Imager (GE Healthcare, Sweden). The maximal absorbance and emission spectra of the Cy-dyes are presented in figure 10.6 and the corresponding filters of the scanner for different wavelengths are given in Table 10.16. The imaging software Image Quant (GE Healthcare, Sweden) was used to calculate the number of pixels where the most intense spots should reach maximal 65000 pixels. The analysis of DIGE gels was carried out with the Delta 2D software (Decodon, Greifswald). After scanning, the gels were subjected to a silver staining.

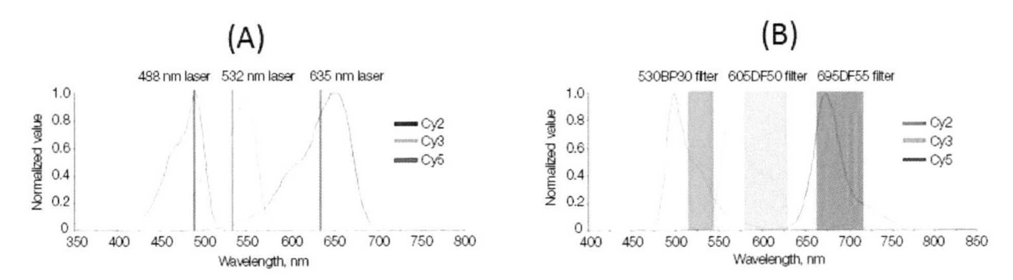

Figure 10.6: Spectra of the Cy-dyes used. (A) Extinction spectra, (B) Emission spectra.

Table 10.16: Extinction and Emission wavelength of the Ettan™ DIGE Imager filters

Dye	Extinction wavelength, [nm]	Emission wavelength, [nm]	Expousure time [ms]
Cy2	480/30	530/40	1.25 - 4
Cy3	540/25	595/25	2 - 4
Cy5	635/30	680/30	3 - 4

10.11.5 GEL STAINING METHODS

Silver staining

For all silver staining steps, 200 mL of each solution were used per gel and incubation was carried out at room temperature on a shaker. First the fixation of the gels took place overnight in fixer solution. On the following day, the solution was carefully poured off and the gels washed twice for 10 min with 30 % (v/v) ethanol. After a washing with MilliQ-H_2O, the gels were covered exactly 1 min with 0.02 % (w/v) thiosulphate. After washing (3x briefly, MilliQ-H_2O), the staining was carried out for 20 min in 0.2 % (w/v) silver nitrate solution with 0.0028 % (v/v) formaldehyde. Subsequently, the gels were washed (3x briefly, MilliQ-H_2O) and finally swung in developer solution until the color development was sufficient. After a final wash with MilliQ-H_2O, the reaction was stopped by addition of the stop solution (50 mM EDTA).

Colloidal coomassie staining

For the colloidal coomassie staining, the gels were first washed twice for 10 min with MilliQ-H_2O to remove the SDS. The staining was carried out overnight in coomassie solution. On the following day the gels were discolored with MilliQ-H_2O.

Table 10.17: Buffers for proteomic analysis

Buffer / Solution	Composition
Fixer solution	40 % (v/v) ethanol
	10 % (v/v) acetic acid in MilliQ-H_2O
Developer solution	3 % (w/v) sodium carbonate
	0.0004 % (w/v) sodium thiosulfate
	0.0185 % (v/v) formaldehyde
Coomassie solution	0.02 % (v/v) CBB-G250
	5 % (w/v) aluminum sulfate-(14-18)-hydrate
	10 % (v/v) ethanol
	2 % (v/v) ortho-phosphoric acid (100 %)

10.11.6 IMAGE ANALYSIS

The analysis of DIGE gels was done with the Delta 2D software (Decodon, Greifswald). The first step was the creation of a new project afterwards the scanned gels were imported into this project and divided into seven groups (standard and samples A to F). The default group included the 12 gels that were scanned with standard Cy2 channel. All other groups were composed of four images each due to the fourfold technical replicates, where two were from Cy3 and two from Cy5 channel.

In the 2D-gel electrophoresis the same protein on different gels does not occur exactly on the same position. To compensate these run differences, a *warping* was performed. This is an image processing method, in which the same protein spots on different gels are superimposed with the help of vectors drawn digitally. Through the predefinition of a warping strategy (*In-Gel Standard Warping Strategy*), the software made possible the indirect warping of the gels. The standard gel with better resolution was selected and matched with the rest of the gels. Thus the gels of the groups were indirectly adjusted. The next step was to create a fusion image. This is a digitally produced image were all spots of the project are displayed. Thereby, the type *Union* was selected for the creation of the image in which the frequency of spots on the gels will be included.

The fusion image was used as the basis for the subsequent *Spot Detection*, were all protein spots are detected and marked with labels. The generated mask spot was transferred to all gels so that each spot on different gels showed the same designation. In the subsequent quantitative evaluation, the normalization of the spot volumes was carried out automatically by the software. The intensity of each spot was set one part in proportion to the total intensity of all spots of a gel and in the other part in relation to the corresponding spot on the standard gel. The percentage spot volume was represented by the program in so-called *Quantitation Tables*.

A heat map of all the samples was created to get a general overview of the differences in the expression profiles of the samples. The relative spot volumes were logarithmically transformed using the *Euclidean distance* and hierarchically clustered by *average linkage method*. Through the hierarchical classification the proteins were arranged according to their similarity in expression pattern.

For the statistical coverage of the data obtained in the DIGE project, the data was submitted to a t-test and an ANOVA (*ANalysis Of VAriance*) with the implemented software in Delta 2D. These are methods of analysis of variance and seek to clarify whether a change in expression pattern is random or whether the change was induced by the parameters to be investigated. To this end, it will be determine whether the variance between the groups under investigation is greater than the variance within a group. In addition, the program compares the spot volumes. If the differences between the different samples is greater than the variability within the group, it can be assumed that there is a change in the expression of the protein, based on the parameters to be investigated in this project (CO_2 content, transfected cell line). The ANOVA was applied to determine the number of proteins that are significantly altered in expression in the entire project and the t-test to compare two groups. The resultant proteins were identified by mass spectrometry.

10.11.7 MASS SPECTROMETRY

The protein identification using mass spectrometry comprises several steps. After the spots were manually picked up from the gels, a tryptic digestion of the proteins followed. The masses of the resultant specific peptide fragments were determined using mass spectrometry. The number and mass of the protein fragments is protein specific and is referred to as the *peptide mass fingerprint* of a protein which can be compared to a database comparison to identify the desired protein.

Spot preparation

For protein identification, the spots were cut out of dried gels with a scalpel and cut out in the fresh gels with a pipette tip. In both cases, the equipment used was washed with 70 % ethanol and MilliQ-H_2O prior to preparation of the next spot. The spots were isolated from several gels to obtain sufficient material for mass-spectrometric investigation. The gel pieces used for the subsequent steps were kept either in 1.5 mL reaction tubes (Eppendorf, Germany) or in 96-well plates (Rotilabo microtest plates made of polystyrene, Carl Roth GmbH, Germany).

Discoloration of the gel spots

If the silver stained gels were used as starting material for protein identification, the gel pieces had to be discolored before the tryptic digestion. These were covered with sufficient volume of discoloration solution (1:1-mixture of 30 mM potassium hexacyano-ferrate III, and 100 mM sodium thiosulfate), briefly vortexed and incubated on a shaker at room temperature until the silver stain was no longer visible. Subsequently, the gel pieces were washed several times with LiChrosolv-water (Merck, Darmstadt, Germany), to removed the residues of the discoloration.

In-gel tryptic digestion

The proteolytic enzyme trypsin, which splits the protein under slightly alkaline conditions (pH 7-9) at the C-terminus of arginine and lysine, unless proline is the following amino acid, was used for in-gel digestion of proteins.

To create the slightly alkaline conditions, the gel pieces were mixed twice for 10 min each in a sufficient volume of a 30 % (v/v) acetonitrile (ACN) solution in 0.1 M ammonium. Thereafter, the gel pieces were dried for about 30 min in the vacuum centrifuge. Depending on the number and size of pieces they were rehydrated with 1-10 μL of 3 mM Tris-HCl buffer (pH 8.8) with 10 ng/μL trypsin (Sequencing grade modified trypsin, Promega, Mannheim). After an incubation period of 30 min, further 2-20 μL of 3 mM Tris-HCl buffer (pH 8.8) without trypsin were added and the solution incubated overnight. On the following day 2-20 μL of LiChrosolv water were added and incubated for 15 min incubation.Finally it followed the addition of 5-20 μL of 0.1 % (v/v) trifluoroacetic acid (TFA) in 30 % (v/v) ACN. After 20 min shaking at room temperature, the samples were stored at -20°C until use.

MALDI target preparation

The preparation of the target (MTP AnchorChipTM 600/384, Bruker Daltonics, Germany) was carried out in two different ways. First, a C18 ZipTip treatment was carried out and, second, the tryptic digested samples were directly pipetted on the MALDI target. In both cases, a α-cyano-hydroxy-cinnamic acid (HCCA) matrix was initially prepared. For this purpose, place a spatula tip of HCCA into a 1.5 mL reaction vessel and suspended with 1 mL of 50 % (v/v) ACN. Before using the mixture, centrifuge for 5 min at 16200x *g*. In both cases, a MALDI-standard (100 pmol/μL) was applied to the corresponding calibration position of the target. This consisted of the following peptides: angiotensin 2 acetate Human, Substance P, ACTH clip 1-17, ACTHclip 18-39. The standard was diluted 1:10 with 10 % (v/v) ACN/0.5 % (v/v) TFA. A mixture of 0.5 μL of this dilution with 0.5 μL matrix was applied per calibration position. For the ZipTip treatment, a 10 μL pipette with a Zip-Tip pipette tip (Millipore Corporation, USA) was used. First, the ZipTip-tip was rinsed twice with 10 μL 50 % (v/v) ACN and then equilibrated twice with 10 μL of a 0.1 % (v/v) TFA solution. For binding of the peptides, the sample was pipetted 10 times. The washing of the peptides was carried out by pipetting 0.1 % (v/v) TFA solution (5x10 μL). For elution, 5x3 μL HCCA matrix were pipetted and the transferred to the matrix. For the direct preparation, 1.5 μL sample and 1 μL

HCCA matrix were mixed and applied to the target.

MALDI-TOF-MS measurement

Mass spectrometric analysis was performed with an Ultraflex MALDI-TOF mass spectrometer (Bruker Daltonics, Germany). The spectra were taken at a laser intensity of 20 to 30 %. 150-200 shots were summed for the calibration spectra, for samples spectra of 500 shots. The software flexControl² 2.0 (Bruker Daltonics, Germany) was used to control the device.

Data analysis

The spectra were annotated and the corresponding mass lists automatically generated with the software used. In part, the spectra were intern manually re-calibrated using the autolysis peaks of modified trypsin (m/z 842.5094 and m/z 2211.104). The connection of the database search to the MASCOT program (Matrix Science Ltd., UK) was made with the BioToolsTM 2.2 software (Bruker Daltonics, Germany). The search was carried out in the SwissProt/UniProt protein database under stetting of defined search parameters, such as taxonomy (Mammalia), enzyme (trypsin), modifications (carbamidomethylation), oxidation of methionine, max. missing enzymatic cleavage site (1), mass tolerance ($\pm$ 20-150 ppm), mass values ($[M+H^+]$ monoisotopic).

CHAPTER 11

BIBLIOGRAPHY

ABABDATZE-BAVIL A (2004) Convenient and versatile subcellular extraction procedure that facilitates protein expression profiling and functional protein analysis. *Proteomics* **4**:397-405

ABSTON LR AND MILLER WM (2005) Effects of NHE1 expression level on CHO cell responses to environmental stress. *Biotechnol Prog* **21**:562-567

ALBERTS B, BRAY D, LEWIS J, RAFF M, ROBERTS K, WATSON, JD (2002) Molecular Biology of the cell. *Garland Publishing*

ALTAMIRANO C, ILLANES A, BECERRA S, CAIRÓ JJ, GÒDIA F (2006) Considerations on the lactate consumption by CHO cells in the presence of galactose. *J Biotecnol* **125**:547-556

ALTAMIRANO C, PAREDES C, ILLANES A, CAIRÓ JJ, GÒDIA F (2004) Strategies for fed-batch cultivation of t-PA producing CHO cells: substitution of glucose and glutamine and rational design of culture medium. *J Biotechnol* **110**:171-179

ALVAREZ BV, LOISELLE FB, SUPURAN CT, SCHWARTZ GJ, CASEY JR (2003) Direct extracellular interaction between carbonic anhydrase IV and the human NBC1 sodium/bicarbonate co-transporter. *Biochem* **42**:12321-12329

AMLAL H, BURNHAM CE, SOLEIMANI M (1999) Characterization of Na^+/HCO_3^- cotransporter Isoform NBC-3. *Am J Physiol* **276**:F903-F913

ANDERSEN DC, BRIDGES T, GAWLITZEK M, HOY C (2000) Multiple cell culture factors can affect the glycosylation of Asn-184 in CHOproduced tissue-type plasminogen activator. *Biotechnol Bioeng* **70**:25–31

ANDERSEN DC AND GOOCHEE CF (1994) The effect of the cell culture conditions on the oligosaccharide strutures of secreted glycoproteins. *Current Opinion Biotechol* **5**:546-549

ARAI H, KOIZUMI H, AOKI J, INOUE K (2002) Platelet-activating factor acetylhydrolase (Paf-AH). *J Biochem* **131**:635-640

AUNINS JG, GLAZOMITSKY K, BUCKLAND BC (1993) *Cell culture reactor desgin: known and unknown.* In: Nienow AW (ed) Proc. 3rd int. conf. on bioreactor and bioprocess fluid dynamics. Mechanical Engineering Publications Ltd., London, pp 175-190. ISBN 0 85298873 7

BASERGA R (1985) The Biology of Cell Reproduction, Harvard University Press, Cambridge

BERG JM, TYMOCZKO JL, STRYER L (2003) Stryer Biochemie. *Spektrum Akademischer Verlag*, Heidelberg, Berlin, 5th edition

BIRCH JR AND RACHER AJ (2006) Antibody production. *Adv Drug Deliv Rev* **58**:671-685

BIRCH JR (2005) Challenges and opportunities in the large scale production of therapeutic proteins. *Presented at the 19th ESACT Meeting*, Harrogate/U.K., 5th-8th June, 2005

BORYS MC, LINZER DI, PAPOUTSAKIS ET (1993) Culture pH affects expression rates and glycosylation of recombinant mouse placental lactogen proteins by Chinese hamster ovary (CHO) cells. *Biotechnol NY* **11**:720-724

BÜNTEMEYER H (1988) *Entwicklung eines Perfusionssystems zur kontinuierlichen Kultivierung tierischer Zellen in Suspension.* Dissertation, Universität Hannover

BUTLER M (2005) Animal cell cultures: recent achievements and perspectives in the production of biopharmaceuticals. *Appl Microbiol Biotechnol* **68**:283-291

CARVALHAL AVSS (2003) Cell Growth Arrest by Nucleotides, Nucleosides and Bases as a Tool for Improved Production. Poster, P-1.10 *18th ESACT Meeting.* **Tylösand**

CASEY JR (2006) Why bicarbonate? *Biochem Cell Biol* **84**:930-939

CAVET ME, AKTHER S, DE MEDINA FS, DONOWITZ M, TSE CM (1999) Na^+/H^+ exchangers (NHE1-3) have similar turnover numbers but different percentages on the cell surface. *Am J Physiol* **277**:C1111-C1121

CHEN JC AND CHESLER M (1992) pH transients evoked by excitatory synaptic transmission are increased by inhibition of extracellular carbonic anhydrase. *Proc Natl Acad Sci USA* **89**:7786-7790

CHEN Q AND ANDERSON DR (1997) Effects of CO_2 on intracellular pH and contraction of retinal capillary pericytes. *Investigative ophthalmology and visual science* **38**: 643-651

CHOW S AND HEDLEY D (1997) Flow Cytometric Measurement of Intracellular pH. *Curr Prot in Cytometry* Unit 9.3.1 - 9.3.10, John Wiley and Sons, Inc

CHOW S, HEDLEY D, TANNACOK I (1996) Flow cytometric calibration of intracellular pH measurements in viable cells using mixtures of weak acids and bases. *Cytometry* **24**:360-367

CHU L AND ROBINSON DK (2001) Industrial choices for protein production by large-scale cell culture. *Curr. Opin. Biotechnol.* **12**:180-187

CORENA MDP, SERON TJ, LEHMAN HK, OCHRIETOR JD, KOHN A, TU C, LINSER PJ (2002) Carbonic anhydrase in the midgut of larval *Aedes aegypti*: cloning, localization and inhibition. *J Exp Biol* **205**:591-602

COUNILLON L AND POUYSSEGUR J (1993) Nucleotide sequence of the Chinese hamster Na^+/H^+ exchanger NHE-1. *Biochim Biophys Acta* **1172**:343-345

COUNILLON L AND POUYSSEGUR J (2000) The expanding family of eucaryotic Na^+/H^+ exchangers. *J. Biol. Chem.* **275**: 1-4

COUNILLON L, SCHOLZ W, LANG HL, POUYSSÉGUR J (1993) Pharmacological characterization of stably transfected Na^+/H^+ antiporter isoforms using amiloride analogs and a new inhibitor exhibiting anti-ischemic properties. *Mol Pharmacol* **44**:1041-1045

DANCKWERTS PV (1970) *Gas-liquid reactions.* McGraw Hill, New York

DAVENPORT HW (1939) Gastric carbonic anhydrase. *J Physiol* **97**:32-43

DE ZENGOTITA VM, SCHMELZER AE, MILLER WM (2002) Characterisation of hybridoma cell responses to elevated pCO_2 and osmolality: intracellular pH, cell size, apoptosis and metabolism. *Biotechnol Bioeng* **77**:369-380

DECKWER WD (1985) *Reaktionstechnik in Blasensäulen.* Otto Salle Verlag GmbH and Co. KG, Frankfurt am Main, 63-72

DOBBYN HC, HILL K, HAMILTON TL, SPRIGGS KA, PICKERING BM, COLDWELL MJ, DE MOOR CH, BUSHELL M, WILLIS AE (2008) Regulation of BAG-1-IRES-mediated translation following chemotaxis stress. *Oncogene* **27**:1167-1174

DOLZ M, O'CONNOR J-E, LEQUERICA JL (2004) Flow cytometric kinetic assay of the activity of Na^+/H^+ antiporter in mammalian cells. *Cytometry* **61A**:99-104

EIBL, R.; EIBL, D.; PÖRTNER, R.; CATAPANO, G.; CZERMAK, P. (2008) *Cell and Tissue Rection Engineering. Series: Principles and Practice.* Springer Berlin Heidelberg

EISNER, DA, KENNING NA, O'NEILL SC, POCOCK G, RICHARDS CD, VALDEOLMILLIS M (1989) A novel method for absolute calibration of intracellular pH indicators. *Pfluegers Arch* **413**:553-558

ELLEN T, KE Q, ZHANG P, COSTA M (2008) NDRG1, a growth and cancer realted gene:regulation of gene expression and function in normal and disease states. *Carcionogenesis* **29**:2-8

FEIGE U, MORIMOTO RI, YAHARA I, POLLA BS (1996) Stress-inducible cellular responses. *Birkäuser Verlag*, Basel

FISHER Z, PRADA JAH, TU C, DUDA D, YOSHIOKA C, AN H, GOVINDASAMY L, SILVERMAN D, MCKENNA R (2005) Structural and Kinetic Characterization of Active-Site Histidine as a Proton Shuttle in Catalysis by Human Carbonic Anhydrase II *Biochem* **44**:1097-1105

FITZPATRICK L, JENKINS HA, BUTLER M (1993) Glucose and glutamine metabolism of a murine B-Lymphocyte hybridoma grown in batch culture. *Appl Biochem Biotechnol* **43**:93-116

FRELIN C, VIGNE P, LADOUX A, LAZDUNSKI M (1988) The regulation of the intracellular pH in cells from vertebrates. *Eur J Biochem* **174**:3-14

GARNIER A, VOYER R, TOM R, PERRET S, JARDIN B, KAMEN A (1996) Dissolved carbon dioxide accumulation in a large scale and high density production of TGFβ receptor with baculovirus infected Sf-9 cells. *Cytotechnology* **22**:53-63

GELFAND DH, STOFFEL S, LAWYER FC, SAIKI RK (1989) Purified Thermostable Enzyme.*United States patent number 4,889,818,*. December 26, 1989

GENIS C (2007) X-ray crystallographic studies of human carbonic anhydrase II captures the intermediate of the hydrolysis of thioxolone. *Journal of Undergraduate Research* **9**

GOUDAR CT, MATANGUIHAN R, LONG E, CRUZ C, ZHANG C, PIERT JM, KONSTATONIV KB (2007) Decreased pCO_2 accumulation by eliminating bicarbonate addition to high cell-densitiy cultures. *Biotechnol Bioeng* **96**:1107-1117

GRAY DR, CHEN S, HOWARTH W, INLOW D, MAIORELLA BL (1996) CO_2 in large-scale and high-density CHO cell perfusion culture. *Cytotechnology* **22**:65-78

GROSS E, PUSHKIN A, ABULADZE N, FEDOTOFF O, KURTZ I (2002) Regulation of the sodium bicarbonate cotransporter kNBC1 function: role of Asp(986), Asp(988) and kNBC1-carbonic anhydrase II binding. *J Physiol* **544**:679-685

GUDERMANN F, ZIEMECK P, LEHMANN J (1997) CeDeX: Automated Cell Density Determination. In: Carrondo MJT, Griffith B, Moreira JLP (eds) *Animal Cell Technology from Vaccines to Genetic Medicine*, Kluwer Academic Publishers, Dordrecht 301-307

GUYTON AC (1991) *Textbook of medical physiology.* Philadelphia: W.B. Saunders Company

HAAS J, PARK EC, SEED B (1996) Codon usage limitation in the expression of HIV-1 envelope glycoprotein. *Curr Biol* **6**:315-324

HANSEN H AND EMBORG C (1994) Influence of ammonium on growth, metabolism, and productivity of a continuous suspension Chinese hamster ovary cell culture. *Biotechnol Prog* **10**:121-124

HEGARDT, F.G. (1999) Mitochondrial 3-hydroxy-methylglutaryl-CoA synthase: a controlenzyme in ketogenesis. *Biochem J* **338**:569-582

HIGBIE R (1935) The rate of absorption of a pure gas into a still liquid during short periods of exposure. *Trans AIChE*, **31**:365-389

ISHIMI (1997) A DNA helicase activity is associated with an mcm 4, -6 and -7 protein complex. *J Biol Chem* **272**:24508-24513

ISHIZAKI A, SHIBAI H, HIROSE Y, SHIRO T (1971). Dissolution and dissociation of carbon dioxide in the model system. Part I: Studies on the ventilation in submerged fermentations. *Agric Biol Chem*, **35**:1733-1740

JACKSON RJ, HOWELL MT, KAMINSKI A (1990) The novel mechanism of initiation of picornavirus RNA translation. *Trends Biochem Sci* **15**:477-483

JANG SK, KRAUSSLICH HG, NICKLIN MJ, DUKE GM, PALMENBERG AC, WIMMER E (1988) A Segment of the 5' nontranslated region of encephalomyocarditis virus RNA directs internal entry of ribosomes during in vitro translation. *J Virol* **62**:2636-2643

JONES RP AND GREENFIELD PF (1982) Effect of carbon dioxide on yeast growth and fermentation. *Enzyme Microb Technol*, **4**:210-223

JUNG H, WANG SY, YANG IW, HSUEH DW, YANG WJ, WANG TH, WANG HS (2003) Detection and treatment of mycoplasma contamination in cultured cells. *Chang Gung Med J* **26**:250-258

KAGEMANN G, HENRICH B, KUHN M, KLEINERT H, SCHNORR O (2005) Impact of Mycoplasma hyorhinis infection on L-arginine metabolism: differential regulation of the human and murine iNOS gene *Biol Chem* **386**:1055-1063

KAHLIFAH RG (1971) The carbon dioxide hydration activity of carbonic anhydrase. I. Stop-flow kinetic studies on the native human isoenzymes B and C. *J Biol Chem* **246**:2561-2573

KAHN AM, CRAGOE EJ JR, ALLEN JC, HALLIGAN RD, SHELAT H (1990) Na^+-H^+ and Na^+-dependent Cl^--HCO_3^- exchange control pH_i in vascular smooth muscle. *Am J Physiol* **259**:C134-C143

KARMAZYN M, SOSTARIC JV, GAN XT (2001) The myocardial Na^+/H^+ exchanger: a potential therapeutic target for the prevention of myocardial ischaemic and reperfusion injury and attenuation of postinfarction heart failure. *Drugs* **61**:375-389

KIMURA R AND MILLER WM (1997) Glycosylation of CHo-derived recombinant tPA produced under elevated pCO_2. *Biotechnol Prog* **13**:311-317

KRAPF R, BERRY CA, ALPEM RJ, RECTOR, FC (1988) Regulation of cell pH by ambient bicarbonate, carbon dioxide tension, and pH in the rabbit proximal convoluted tubule. *J Clin Invest* **81**:381-389

KUBBIES M, GOLLER B, GIESE G (1996). High-resolution cell cycle analysis of cell cycle-regulated gene expression. In Flow cytometry applications in cell culture. M. Al-Rubeai and A. N. Emery. New York, Marcel Dekker

KUMAR R, CONKLIN DS, MITTAL V (2003) High-throughput selection of effective RNAi probes for gene silencing *Genome Research* **13**:2333-2340

KURANO N, LEIST C, MESSI F, KURANO S, FIECHTER A (1990) Growth behaviour of Chinese hamster ovary cells in a compact loop bioreactor. Part 2. Effects of medium components and waste products. *J Biotecnol* **15**:113-128

LANDESMAN-BOLLAG E, ROMIEU-MOUREZ R, SONG DH, SONENSHEIN GE, CARDIFF RD, SELDIN DC (2001) Protein kinase CK2 in mammary gland tumorigenesis. *Oncogene* **20**:3247-3257

LEELAVATCHARAMAS V, EMERY AN, AL-RUBEAI M (1996) Monitoring the proliferative capacity of cultured animal cells by cell cycle analysis. In Flow cytometry applications in cell culture. M. Al-Rubeai and A. N. Emery. New York, Marcel Dekker

LEHNINGER AL (1982) *Principles of biochemistry.* Worth Publishers, Inc., New York

LEWIS WK AND WHITMAN WG (1924) The two-film theory of gas absorption. *Ind Eng Chem*, **16**(12):1215-1239

LI X, ALVAREZ B, CASEY JR, REITHMEIER RAF, FLIEGEL L (2002) Carbonic Anhydrase II binds to and enhances activity of the Na^+/H^+ exchanger *J Biol Chem* **277**:36085-36091

LI X, LIU Y, ALVAREZ BV, CASEY JR, FLIEGEL L A novel carbonic anhydrase II binding site regulates NHE1 activity (2006) *Biochemistry* **45**:2414-2424

LINCOLN CK AND GABRIDGE MG (1998) Cell culture contamination: sources, consequences, prevention, and elimination. *Methods in Cell Biology* **57**:49-65

LINDSEY AE, SCHNEIDER K, SIMMONS DM, BARON R, LEE BS, KOPITO RR (1990) Functional expression and subcellular localization of an anion exchanger cloned from Choroid Plexus. *Proc Natl Acad Sci USA* **87**:5278-5282

LIU W, XIONG Y, GOSSEN M (2006) Stability and homogeneity of transgene expression in isogenic cells. *J Mol Med* **84**:57-64

LLORCA O, MARTÍN-BENITO J, GRANTHAM J, RITCO-VONSOVICI M, WILLISON KR CARRASCOSA JL, VALPUESTA JM (2001) The sequential allosteric ring mechanism in the eukaryotic chaperonin-assisted folding of actin and tubulin. *EMBO J* **20**:4065-4075

LU J, DALY CM, PARKER MD, GILL HS, PIERMARINI PM, PELLETIER MF, BORON WF (2006) Effect of human carbonic anhydrase II on the activity of the human electrogenic Na^+/HCO_3^- cotransporter NBCe1-A in Xenopus oocytes. *J Biol Chem* **281**:19241-19250

MADSHUS IH (1988) Regulation of intracellular pH in eukaryotic cells. *Biochem J* **250**:1-8

MAIER U AND BÜCHS J (2001) Characterization of the gas-liquid mass transfer in shaking bioreactors. *Biochem Eng J*, **7**:99-106 MAREN TH (1967) Carbonic anhydrase: chemistry, physiology, and inhibition. *Physiol Rev* **47**:595-781

MASON MJ, SMITH JD, GARCIA-SOTO JJ, GRINSTEIN S (1989) Internal pH-sensitive site couples Cl^-/HCO_3^- exchange to Na^+/H^+ antiport in lymphocytes. *Am J Physiol* **256**:C428-C433

MATANGUIHAN R, SAJAN E, ZACHARIOU M, OLSON C, MICHAELS J, THRIFT J, KONSTATINOV K (2001) Solution to the high dissolved CO_2 problem in high-density perfusion culture of mammalian cells. In *Animal Cell Culture Technology: From target to market.* Kluwer, The Netherlands:399-402

MATZ MV, FRADKOV AF, LABAS YA, SAVITSKY AP, ZARAISKY AG, MARKELOV ML, LUKYANOV SA (1999) Fluorescent proteins from nonbioluminescent Anthozoa species. *Nature Biotechnol* **17**:969-973

MEIER SJ (2005) *Cell culture scale-up: mixing, mass transfer, and use of appropriate scale-down models.* Biochemical engineering XIV. Harrison Hot Springs, Canada

MELDRUM NU AND ROUGHTON FJW (1932) Some properties of carbonic anyhrase, the CO_2 enzyme present in blood. *J Physiol* **75**:15-16

MELDRUM NU AND ROUGHTON FJW (1933) Carbonic anhydrase: Its preparation and properties. *J Physiol* **80**:113-142

MERTEN O-W (2006) Introduction to animal cell culture technology - past, present and future. *Cytotechnology* **50**:1-7

MIRJALILI A, PARMOOR E, MORADI BS, SARKARI B (2005) Microbial contamination of cell cultures: a 2-years study *Biologicals* **33**:81-85

MITZ MA (1979) Carbon dioxide biodynamics: a new concept of cellular control. *Journal of Theoretical Biologie*, **80**(4):537-351

MOHAN C, KIM YG, KOO J, LEE GM (2008) Assessment of cell engineering strategies for improved therapeutic protein production in CHO cells. *Biotecnol J* **3**:624-30

MOSTAFA SS AND GU X (2003) Strategies for improved dCO_2 removal in large-scale fed-batch cultures. *Biotecnol Prog* **19**:45-51

MUTZALL K (1993) *Einführung in die Fermentationstechnik.* Behrs Verlag GmbH, Hamburg

NAIR SK AND CHRISTIANSON DW (1991) Unexpected pH-dependent conformation of His64, the proton shuttle of carbonic anhydrase II. *J AM Chem Soc* **113**:9455-9458

NAKAMURA N, TANAKA S, TEKO Y, MITSUI K, KANAZAWA H (2005) Four Na^+/H^+ exchanger isoforms are distributed to Golgi and post-Golgi compartments and are involved in organelle pH regulation. *J Biol Chem* **280**:1561-1572

NG HH AND BIRD A (1999) DNA methylation and chromatin modification. *Curr Opin Genet Dev* **9**:158-163

NI M AND LEE AS (2007) ER chaperons in mammalian development and human diseases. *FEBS Letters* **581**:3641-3651

NIENOW AW, LANGHEINRICH C, STEVENSON NC, EMERY AN, CLAYTON TM, SLATER NKH (1996) Homogeneisation and oxygen transfer rates in large agitated and sparged animal cell bioreactors: Some implications for growth and production. *Cytotechnology* **22**:87-94

NIENOW AW (2006) Reactor engineering in large scale animal cell culture. *Cytotechnology* **50**:9-33

NISSOM PM, SANNY A, KOK YJ, HIANG YT, CHUAH SH, SHING TK, LEE YY, WONG KT, HU WS, SIM MY, PHILP R. (2006) Transcriptome and proteome profiling to understanding the biology of high productivity CHO cells. *Mol Biotechnol* **34**:125-140

OH S, CHUA FKF, AL.RUBEAI M (1996) *Flow cytometry applications*

ONKEN U AND LIEFKE E (1989) Effect of total and partial pressure (oxygen and carbon dioxide) on aerobic microbial processes. *Adv Biochem Eng Biotechnol*, **40**:138-169

OREGON STATE UNIVERSITY, OSU (2009) *Flow Cytometry - How Does It Work?* in http://www.unsolvedmysteries.oregonstate.edu, Oregon State University: Environmental Health Sciences Center (EHSC)

ORLOWSKI J AND GRINSTEIN S (1997) Na^+/H^+ exchangers of mammalian cells. *J Biol Chem* **272**:22373-22376

ORLOWSKI J AND GRINSTEIN S (2004) Diversity of the mammalian sodium/proton exchanger SLC9 gene family. *Pfluegers Arch* **447**:549-565

OZTURK SS (1996) Engineering challenges in high-density cell culture systems. *Cytotechnology* **22**:3-16

OZTURK, SS, JORJANI P, TATICEK B, LOWE S, SHACKLEFORD D, LADEHOFF-GUILES D, THRIFT J, BLACKIE J, NAVEH D (1997) Kinetics of glucose metabolism and utilization of lactate in mammalian cell cultures. In: Carrondo MJT (Ed.), *Animal Cell Technology*. Kluwer Academic Publishing, Netherlands:355-360

PANG T, SU X, WAKABAYASHI S, SHIGEKAWA M (2001) Calcineurin homologous protein as an essential cofactor for Na^+/H^+ exchangers. *J Biol Chem* **276**:17367-17372

PARKER MD, MUSA-AZIZ R, ROJAS JD, CHOI I, DALY CM, BORON WF (2008) Characterization of Human SLC4A10 as an Electroneutral Na^+/HCO_3^- Cotransporter (NBCn2) with Cl^- Self-exchange Activity *J Biol Chem* **283**:12777-12788

PASCOE D, ARNOTT D, PAPOUTSAKIS ET, MILLER WM, ANDERSEN DC (2007) Proteome analysis of antibody-producing CHO cell lines with different metabolic profiles. *Biotechnol Bioeng* **98**:391-410

PATTISON RN, SWAMY J, MENDENHALL B, HWANG C, FROHLICH BT (2001) Measurement and control of dissolved carbon dioxide in mammalian cell culture process using an in situ fiber optic chemical sensor. *Biotechnol Prog* **16**:769-774

PAXTON J (1974) PhD thesis, University of Queensland PURKERSON JM AND SCHWARTZ GJ (2007) The role of carbonic anhydrases in renal physiology. *Kidney International* **71**:103-115 PUTNEY LBD (2003) Na-H Exchange-dependent increase in intracellular pH times G2/M entry and transition. *J Biol Chem* **278**:44645-44649

REINERTSEN KV, TONNESSEN TI, JACOBSEN J, SANDVIG K, OLSNES S (1988) Role of chloride/bicarbonate antiport in the control of cytosolic pH; cell-line differences in activity and regulation of antiport. *Biol Chem* **263**:11117-11125

RO H AND CARSON JH (2004) pH microdomains in Oligodendrocytes *J Biol Chem* **279**:37115-37123

ROOS A AND BORON WF (1981) Intracellular pH. *Physiol Rev* **61**:296-434

ROTHMAN RJ, WARREN L, VLIEGENTHART JF, HARD KJ (1989) Clonal analysis of the glycosylation of immunoglobulin G secreted by murine hybridomas. *Biochem* **28**:1377-1384

ROYCE PNC AND THORNHILL NF (1991) Estimation of dissolved carbon dioxide concentrations in aerobic fermentations. *AIChE Journal* **37**:1680-1686

SAMBROOK J AND RUSSELL DW (2001) Molecular Cloning: A Laboratory Manual. *Cold Spring Harbor Laboratory*, 3rd edition

SCHMIDT H-M, ZUMBANSEN M, WITTIG R, BLAICH S, BROWN L, LYER S, POUSTKA A, MOLLENHAUER J, NIX M (2004) Use of Nucleofector® technology to establish stably expressing cell lines. *Technol Note*, Amaxa Biosystems

SERRATO JA, HERNÁNDEZ V, ESTRADA-MONDACA S, PALOMARES LA, RAMIREZ OT (2007) Differences in the glycosylation profile of a monoclonal antibody produced by hybridomas cultured in serum-supplemented, serum-free or chemically defined media. *Biotechnol Appl Biochem* **47**:113-124

SETH G, CHARANIYAA S, WLASCHINA KF, HUA W-S (2007) In pursuit of a super producer-alternative paths to high producing recombinant mammalian cells. *Curr Opinion Biotechnol* **18**:557-564

SILVA NLCL, HAWORTH RS, SINGH D, FLIEGEL L (1995) The carboxyl-terminal region of the Na^+/H^+ exchanger interacts with mammalian heat shock protein. *Biochem* **34**:10412-10420

SILVERMAN DN (1982) Carbonic anhydrase: Oxygen-18 exchange catalyzed by an enzyme with rate-contributing proton-transfer steps. *Methods Enzymol* **87**:732-752

SLEPKOV ER, RAINEY JK, SYKES BD, FLIEGEL L (2007) Structural and functional analysis of Na^+/H^+ exchanger. *Biochem J* **401**: 623-633

SLY WS AND HU PY (1995) Human Carbonic Anhydrases and Carbonic Anhydrase Deficiencies. *Annu Rev Biochem* **64**: 375-601

SLY WS, WHYTE MP, SUNDARAM V, TASHIN RE, HEWETT-EMMETT D, GUIBAUD P, VAINSEL M, BALUARTE HJ, GRUSHKIN A, AL-MOSAWI M, SAKATI N, OHLSSON A (1985) Carbonic anhydrase II deficiency in 12 families with the autosomal recessive syndrome of osteopetrosis with renal tubular acidosis and cerebral calcification. *N Engl J Med* **313**:139-145

SOUTHERN PJ AND BERG P (1982) Transformation of mammalian cells to antibiotic resistance with a bacterial gene under control of the SV40 early region promoter. *J Mol Appl Genet* **1**:327-341

SPITZER KW, SKOLNICK RL, PEERCY BE, KEENER JP, VAUGHAN-JONES RD (2002) Facilitation of intracellular H^+ ion mobility by CO_3/HCO_3^- in rabbit ventricular myocytes is regulated by carbonic anhydrase. *J Physiol* **541**:159-167

SRERE PA (1987) Complexes of sequential metabolic enzymes. *Annu Rev Biochem* **56**:89-124

STEINER H, JONSSON B-H, LINDSKOG S (1975) The catalytic mechanism of carbonic anhydrase. Hydrogen-isotope effects on the kinetic parameters of the human C isoenzyme. *Eur J Biochem* **59**:253-259

STERLING D AND CASEY JR (1999) Transport activity of AE3 chloride/bicarbonate anion-exchange proteins and their regulation by intracellular pH. *Biochem J* **344**:221-229

STERLING D, REITHMEIER RA, CASEY JR (2001) A transport metabolon. Functional interaction of carbonic anhydrase II and chloride/bicarbonate exchangers. *J Biol Chem* **276**:47886-47894

SZATKOWSKI MS AND THOMAS RC (1986) New method for calculating pHi from accurately measured changes in pHi induced by a weak acid and base. *Pfluegers Arch* **107**:59-53

TATICEK R, PETERSEN S, KONSTATINOV K, NAVEH D (1998) Effect of dissolved carbon dioxide and bicarbonate on mammalian cell metabolism and recombinant protein productivity in high density perfusion culture. *ACS Annual Meeting, San Diego, CA*

THOMAS RC (1989) Bicarbonate and pH_i response. *Nature* **337**:601

THOMAS RCJ (1977) The role of bicarbonate, chloride and sodium ions in the regulation of intracellular pH in snail neurones. *J Physiol Lond* **273**:317-338

TONG S-K, BRION LP, SUAREZ C, CHESLER M (2000) Interstitial Carbonic Anhydrase (CA) Activity in Brain Is Attributable to Membrane-Bound CA Type IV. *The Journal of Neuroscience* **20**:8247-8253

TONNESSEN TI, SANDVIG K, OLSNES S (1990) Role of Na^+-H^+ and Cl^--HCO_3^- antiports in the regulation of cytosolic pH near neutrality. *Am J Physiol* **258**:C1117-C1126

TRAMPER J (1995) Oxygen gradients in animal-cell bioreactors. *Cytotechnology* **18**:27-34

VANDENBERG JI, METCALFE JC, GRACE AA (1993) Mechanisms of pH_i recovery after global ischemia in the perfused heart. *Circ Res* **72**:993-1003

VAUGHAN-JONES RD (1982) *Intracellular pHi: Its measurement, regulation, and utilization in cellular functions.* Edited by Nuccitelli R, Deamer DW, New York, Alan R. Liss Inc:239-252

VERKMAN AS AND ALPERN RJ (1987) Kinetic transport model for cellular regulation of pH and solute concentration in the renal proximal tubule. *Biophys J* **51**:533-546

VINCE JW AND REITHMEIER RA (1998) Carbonic anhydrase II binds to the carboxyl-terminus of human band 3, the erythrocyte Cl^-/HCO_3^- exchanger. *J Biol Chem* **273**:28430-28437

VINCE JW AND REITHMEIER RA (2000) Identification of the carbonic anhydrase II binding site in the Cl^-/HCO_3^- anion exchanger AE1. *Biochem* **39**:5527-5533

VINCENT SH AND SILVERMAN DN (1982) Carbonic anhydrase activity in Mitochondria from rat liver. *J Biol Chem* **257**:6850-6855

VOET D AND VOET JD (1995) Biochemistry, 2nd Ed., *Wiley and Sons*, New York

VON GROLL A, LEVIN Y, BARBOSA MC, RAVAZZOLO AP (2006) Linear DNA low efficiency transfection by liposome can be improved by the use of cationic lipid as charge neutralizer. *Biotechnol Prog* **22**:1220-1224

VOOS W (2003) A new Connection: Chaperones meet a mitochondrial receptor. *Molecular Cell* **11**:1-10

WAKABAYASHI S, FAFOURNOUX P, SARDET C, POUYSSEGUR J (1992) The Na^+/H^+ antiporter cytoplasmic domain mediates growth factor signals and controls 'H^+-sensing'. *Proc Natl Acad Sci USA* **89**:2424-2428

WAKABAYASHI S, IKEDA T, IWAMOTO T, POUYSSEGUR J, SHIGEKAWA M (1997) Calmodulin-binding autoinhibitory domain controls "pH-sensing" in the Na^+/H^+ exchanger BHE1 through sequence specific interaction. *Biochem* **36**:12854-12861

WALSH, G (2009) *Post-translational modification of protein biopharmaceuticals*, Wiley-Blackwell

WELSH JP AND AL-RUBEAI M (1996) The relationship between intracellular pH and cell cycle in cultured animal cells using SNARF-1 indicator. *Flow Cytometry: Application in cell culture.* New York, Marcel Dekker Inc:163-175

WENGER SL, SENFT JR, SARGENT LM, BAMEZAI R, BAIRWA N, GRANT SG (2004) Comparison of established cell lines at different passages by karyotype and comparative genomic hybridization. *Biosci. Rep.* **24**: 631-639

WHITELAW E, SUTHERLAND H, KEARNES M, MORGAN H, WEAVING L, GARRICK D (2001) Epigenetic effects on transgene expression. *Methods Mol Biol* **158**:351-368

WICKNER W AND SCHEKMAN R (2008) Membrane fusion. *Ant Struc Mol Biol* **15**:658-664

WIEDER ED, HANG H, FOX MH (1993) Measurement of intracellular pH using flow cytometry with carboxy-SNARF-1. *Cytometry* **14**:916-921

WINGENS M (2008) Differentielle Proteomanalyse in der Zellkulturtechnik: Methodenentwicklung und Anwendungen.*PhD Dissertation*, University of Bielefeld

WU P, RAY NG, SCHULER ML (1993) A computer model for intracellular pH regulation in CHO. *Biotechnol Prog* **9**:374

WURM FM, GWINN KA, KINGSTON RE (1986) Inducible overproduction of the mouse c-myc protein in mammalian cells. *Proceedings of the National Academy of Sciences of the USA* **83**:5414-5418

YAO H, MA E, GU X-Q, HADDAD GG (1999) Intracellular pH regulation of CA1 neurons in Na^+/H^+ isoform 1 mutant mice. *J Clin Invest* **104**:637-645

ZANGHI JA, SCHMELZER AE, MENDOZA TP, KNOP RH, MILLER WM (1999) Bicarbonate concentration and osmolality are key determinant in the inhibition of CHO cell polysialylation under elevated pCO_2 or pH. *Biotechnol Bioeng* **65**:182-191

ZHANG Y, CHERNOVA MN, STUART-TILLEY AK, JIANG L, ALPER SL (1996) The cytoplasmic and transmembrane domains of AE2 both contribute to regulation of anion exchange by pH. *J Biol Chem* **271**:5741-5749

ZHU MM, GOYAL A, RANK DL, GUPTA SK, BOOM TV, LEE SS (2005) Effects of elevated pCO_2 and osmolality on growth of CHO cells and production of antibody-fusion protein B1: A case study. *Biotechnol Prog* **21**:70-77

Introduction to Flow Cytometry - A Learning Guide, Manual Part Number 11-11032-01. (2000) *BD Biosciences*

Living Colors® pIRES2-ZsGreen1 Vector (2005) *Clontechniques*, 18

Living Colors® User Manual Volume II: Reef Coral Fluorescent Proteins (2003) *Clontech Laboratories Inc.* Protocol n°. PT3404-1, Version n°. PR37085

Operations manual YSI Biovision 8500: CO_2 monitor. (2002) *Yellow Springs* USA

pIRES2-ZsGreen1 vector information. (2005) *Clontech Laboratories Inc.* Protocol n° PT3824-5, Version n° PR52081

Chapter 12

Nomenclature

Symbol	Description	Units
Amino acids		
Ala	Alanine	-
Arg	Arginine	-
Asn	Asparagine	-
Asp	Aspartic acid	-
Cys	Cysteine	-
Gln	Glutamine	-
Glu	Glutamic acid	-
Gly	Glycine	-
His	Histidine	-
Ile	Isoleucine	-
Leu	Leucine	-
Lys	Lysine	-
Met	Methionine	-
Phe	Phenylalanine	-
Pro	Proline	-
Ser	Serine	-
Thr	Threonine	-
Trp	Tryptophan	-
Tyr	Tyrosine	-

Symbol	Description	Units
Val	Valine	-
Nucleobases		
A	Adenine	-
C	Cytosine	-
G	Guanine	-
T	Thymine	
Abreviations		
a	specific mass transfer area	$\left[\frac{m^2}{m^3}\right]$
ACN	acetonitrile	-
Act_{smpl}	activity of CAII in the sample	-
Act_{std}	activity of CAII in the standard	-
ACTZ	acetazolamide	-
Amp^r	ampicillin resistance	-
ANOVA	Analysis of variance	-
BA	butyric acid	-
BCA	bicinchoninic acid	-
c_i	concentration of the component "i"	$\left[\frac{mol}{L}\right]$
Cedex	Cell Density Examination	-
CHO	Chinese Hamster Ovary	-
CMV	cytomegalovirus	-
CO_2	Carbon Dioxide	-
DHFR	Dihydrofolate Reductase	-
DMEM	Dulbecco's Modified Eagle Medium	-
DMSO	Dimethyl sulfoxide	-
DNA	DeoxyriboNucleic Acid	-
DNase	DNA nuclease	-
ds	Double Strand	-
E. coli	*Escherichia coli*	-
ECL	Enhanced Chemiluminecence	-
ECMV	encephalomyocarditis virus	-
EtBr	Ethidium Bromide	-

Symbol	Description	Units
EZA	ethoxyzolamide	-
FBS	Fetal bovine serum	-
G_0	cell cycle G0 phase	-
G_1	cell cycle G1 phase	-
G_2	cell cycle G2 phase	-
G0	agalactosylated structure	-
G1	monogalactosylated structure	-
G2	digalactosylated structure	-
G418	Geneticin	-
GOI	Gene of Interest	-
h	hours	-
hCAII	human carbonic anhydrase II	-
HCl	Hydrochloric acid	-
HDFBS	HEPES buffer containing FBS	-
HRP	Horseradish Peroxidase	-
HTPS	Hydroxypyrene trisulfonic acid	-
I_i	maximum emission intensity of component "i"	-
IEF	isoelectric focusing	-
IgG	Immunoglobulin G	-
IRES	Internal Ribosome Entry Site	-
k_H	Henry coefficient of carbon dioxide	$\left[\frac{atm \cdot L}{mol}\right]$
k_L	liquid-phase mass transfer coefficient	$\left[\frac{m}{h}\right]$
LB	Luria-Bertani	-
m_i	weight of component *i*	[g]
MCS	Multiple Cloning Site	-
min	minutes	-
mRNA	Messenger Ribonucleic Acid	-
MTX	Methotrexate	-
NaOH	Natrium hydroxide	-
NEB	New England Biolabs	-
Neo^r	neomycin-resistance gene	-
O_2	Oxygen	-
OD	Optical Density	[-]
OPA	ortho Phthaldialdehyd	-

Symbol	Description	Units
p_{CO_2}	partial pressure of CO_2	[mmHg]
P_t	total pressure	[mmHg]
P	product concentration	$\left[\frac{g_P}{L}\right]$
$P_{CMV\ IE}$	immediate early promoter of cytomegalovirus	-
PI	Propidium iodide	-
PBS	Phosphate Saline Buffer	-
PCR	Polymerase Chain Reaction	-
pH	pH value	-
pH_i	intracellular pH value	-
PI	Propidium Iodide	-
PMSF	phenylmethylsulfonyl fluoride	-
PVDF	polyvinylidene difluoride	-
q_P	Specific Productivity	$\left[\frac{pg_P}{cell \cdot day}\right]$
RQ	respiratory quocient	$\left[\frac{mol\ CO_2}{mol\ O_2}\right]$
RNA	RiboNucleic Acid	-
RNase	RNA nuclease	-
rRNA	Ribosomal RNA	-
RT	room temperature	-
S	cell cycle S phase	-
S	substrate concentration	$\left[\frac{g_S}{L}\right]$
sec	seconds	-
SOB	Super Optimal Broth	-
SOC	Super Optimal Broth with Catabolite repression	-
sp.	species	-
ss	Single Strand	-
T_A	Annealing Temperature	[°C]
TFA	trifluoroacetic acid	-
T_M	Melting Temperature	[°C]
TMA	trimethylamine	-
T	temperature	[K]
t	time	[h]
U.T.	untranslated	-
V	Viability	[%]

Symbol	Description	Units
V_{smpl}	volume of sample used for measurement	-
V_{std}	volume of standard used for measurement	-
VCD	Viable Cell Density	$\left[\frac{cells}{mL}\right]$
WCB	Working Cell Bank	-
X	biomass concentration	$\left[\frac{gx}{L}\right]$
x_{CO2}	concentration of CO_2	$[\%]$
ZsGreen	*Zoanthus* sp. green	-
μ	Specific Growth Rate	$\left[\frac{1}{h}\right]$
μ_{max}	Maximum Specific Growth Rate	$\left[\frac{1}{h}\right]$

Chapter 13

Appendix

Spot ID	Identifiziertes Protein	Synonym	Swiss-Prot		MW	pI	Sequenzabdeckung	MOWSE-Score	Relative Spotvolumina Probe E [%]				MW [%]	Relative Spotvolumina Probe F [%]				MW [%]	Funktionelle Kategorie
			Acc.No.	Spezies	[kDa]	[-]	[%]		Gel4	Gel5	Gel8	Gel12	Probe E	Gel5	Gel6	Gel9	Gel11	Probe F	
AMU_324	Activator of 90 kDa heat shock protein	AHspA1	Q8BK64	Mouse	38,3	5,3	34,3	112,0	75,4	72,6	73,5	77,6	74,8	57,1	55,7	64,7	55,7	58,3	
AMU_575	Endoplasmin precursor	ENPL, GRP94	P08113	Mouse	92,7	4,6	11,3	89,8	45,3	42,5	30,5	64,7	47,0	103,0	122,5	137,1	90,9	113,4	
AMU_709	FK506-binding protein 4	FKBP4	P30416	Mouse	45,9	5,4	31,2	108,0	72,4	78,9	75,0	78,7	76,2	58,1	58,2	65,9	52,2	58,6	
AMU_1037	FK506-binding protein 4	FKBP4, FKBP 52,	P30416	Mouse	45,9	5,4	22,1	87,4	51,9	53,7	55,5	60,6	55,4	83,4	87,6	83,1	73,8	82,0	
AMU_596	78 kDa glucose-regulated protein precursor	GRP78, BiP	P07823	Golden Hamster	72,5	4,9	41,3	278,0	141,6	132,1	120,3	153,3	136,8	245,8	217,5	248,2	248,8	240,1	Stress Proteine/ Chaperone
AMU_604	Stress-70 protein, mitochondrial precurser	GRP75	O35501	Chinese Hamster	73,7	5,8	17,8	110,0	104,1	895,0	110,6	105,2	104,6	139,3	135,9	138,9	135,8	137,5	
AMU_752	Mitochondrial precursor proteins import receptor	TOM70	O94826	Human	67,4	6,9	19,9	67,7	58,6	60,4	70,1	68,6	64,5	113,8	95,4	107,2	112,1	109,3	
AMU_1001	T-complex protein 1 subunit gamm	TCPG	Q6P502	Rat	60,6	6,2	18,5	96,4	68,5	67,9	68,1	62,6	66,8	44,3	45,6	42,9	36,1	42,2	
AMU_783	T-complex protein 1 subunit beta	TCPB	Q5XIM9	Rat	57,5	6,0	35,7	110,0	72,5	65,9	67,4	67,4	68,3	44,5	38,7	39,7	42,4	41,3	
AMU_496	Heat-shock protein beta-1 (HspB1)	HspB1	P15991	Hamster	23,5	6,3	54,0	150,0	43,8	54,7	49,2	53,2	50,2	33,6	41,4	33,4	35,1	36,4	
AMU_665	Actin, cytoplasmic 1	ACTB	P48975	Chinese Hamster	41,7	5,1	24,0	74,2	55,9	64,5	58,0	57,2	58,9	135,8	117,5	121,6	133,7	129,2	
AMU_634	Tubulin beta-5 chain	TBB5	P69893	Chinese Hamster	50,1	4,6	71,4	312,0	108,3	100,1	97,9	105,9	103,0	85,6	82,0	84,9	75,3	82,0	Struktur-proteine
AMU_532	Vimentin	VIME	P02544	Golden	53,7	4,9	47,1	177,0	32,3	45,3	37,4	49,3	41,1	88,8	129,2	185,2	107,6	126,7	
AMU_649	Vimentin	VIME	P02544	Rat	53,7	4,9	44,6	183,0	58,8	73,2	69,9	80,9	70,7	129,4	174,7	138,7	176,7	153,6	
AMU_692	Vimentin	VIME	P02544	Chinese Hamster	53,7	4,8	36,4	140,0	64,6	69,8	42,2	82,0	64,6	166,6	176,2	135,2	156,1	158,6	
AMU_544	Sorting Nexin-4	SNX4	Q91YJ2	Mouse	51,8	5,5	34,4	147,0	52,9	51,0	57,2	57,7	54,7	43,3	39,3	43,7	41,0	41,9	Transport proteine
AMU_809	Sorting Nexin-6	SNX6	Q6P8X1	Mouse	46,6	5,7	19,5	85,9	66,6	72,7	79,0	76,3	75,4	54,3	53,0	50,4	59,3	56,6	
AMU_426	Eukaryotic translation initiation factor 4H	IF4H	Q9WUK2	Mouse	27,3	7,5	46,0	67,6	8,9	12,4	11,9	13,0	11,5	4,1	3,4	3,9	4,4	4,0	
AMU_50	Elongation factor 2	EF2	Q8K164	Rat	95,3	6,4	22,3	108,0	81,8	83,5	75,9	86,7	81,8	69,0	59,1	67,0	61,0	64,0	
AMU_882	Eef2 protein (Fragment)	Eef2	Q8K164	Mouse	33,9	6,4	28,6	73,6	74,0	99,6	103,9	102,8	95,1	160,0	147,1	172,4	157,6	159,5	
AMU_162	Eukaryotic translation initiation factor 3 subunit 2 beta	IF32	P55884	Human	36,9	5,3	29,5	104,0	61,4	62,3	72,1	66,3	65,6	91,6	88,4	100,6	84,2	91,2	Protein-biosynthese
AMU_981	Poly(rC)-binding protein 1	PCBP1	P60335	Mouse	38,0	6,8	30,6	97,4	64,2	72,2	66,7	63,6	67,2	50,6	51,8	53,9	46,9	50,8	
AMU_754	cytopl. Seryl-aminoacyl-tRNA synthetase 1		A2AFS0	Mouse	40,5	5,1	39,5	164,0	86,8	85,0	90,4	93,1	88,8	76,4	72,7	75,1	67,6	73,0	
AMU_432	Heme oxygenase 1	HMOX1	P06762	Rat	33,0	6,1	21,1	62,3	54,6	73,9	73,6	70,8	68,2	37,7	52,3	45,6	44,1	44,9	Redox-kontrolle
AMU_543	cytopl. Hydroxymethyl-glutaryl-CoA synthase	HMCS1	P13704	Chinese Hamster	57,9	5,3	33,8	179,0	56,9	64,6	56,7	60,8	59,7	86,9	85,2	93,1	78,5	86,0	Lipid-metabolismus
AMU_1008	Minichromosome maintenance protein 7	MMP7, mcm 7	Q1PS21	Rat	81,7	5,9	30,9	210,0	60,3	69,3	79,4	79,5	72,1	36,6	39,1	56,7	42,9	43,8	DNA-Replika-tion
AMU_840	Protein NDRG 1	NDRG1	Q6JE36	Rat	43,4	5,7	37,8	91,5	41,1	49,0	47,2	44,1	45,4	5,0	7,3	4,4	4,8	5,4	
AMU_867	Protein NDRG 1	NDRG1	Q6JE36	Rat	43,4	5,7	26,4	71,3	152,9	179,8	211,2	139,7	171,0	64,1	67,6	59,7	57,3	62,2	
AMU_330	NDRG1	NDRG1	Q6JE36	Rat	43,4	5,7	34,0	84,0	104,5	105,5	131,3	108,9	112,6	217,9	189,8	212,2	214,8	208,7	
AMU_329	NDRG1	NDRG1	Q6JE36	Rat	43,4	5,7	25,0	66,0	88,4	87,2	107,9	88,7	93,1	141,7	125,1	120,9	119,9	126,9	
AMU_685	NSFL1 cofactor p47	NSFL1	O35987	Rat	40,7	4,9	44,1	167,0	34,5	42,8	50,1	44,7	43,0	21,2	22,5	19,3	15,1	19,5	
AMU_315	Mammary gland RCB-0527 Jyg-MC cDNA		Q3TLL6	Mouse	41,5	5,2	45,6	172,0	64,2	67,4	87,4	83,8	75,7	129,9	134,4	141,8	94,8	125,2	sonstige
AMU_991	Annexin A1	ANXA1	P07150	Rat	38,8	7,7	16,8	72,2	70,0	68,4	76,7	72,3	69,1	81,3	79,9	81,5	73,4	79,1	
AMU_373	Platelet-activating factor acetylhydrolase IB subunit alpha	LIS1	P43034	Human	46,6	7,2	33,7	86,9	60,2	74,8	76,7	72,3	71,0	112,9	97,1	130,6	105,4	111,5	
AMU_952	Heterogeneous nuclear ribonucleoproteins A2/B1	ROA2	Q2HJ60	Bovine	36,0	9,1	23,5	96,5	45,0	64,1	64,2	64,5	59,4	130,7	144,4	114,9	123,8	123,8	
AMU_313	Heterogenous nuclear ribonucleoprotein F	HNRPF	Q972X1	Mouse	46,0	5,2	40,7	97,3	67,6	67,5	67,0	62,7	66,2	39,3	46,5	48,0	37,8	42,9	mRNA-Prozessie-rung
AMU_311	Heterogenous nuclear ribonucleoprotein F	HNRPF	Q972X1	Mouse	46,0	5,2	43,9	99,9	84,6	83,4	72,6	79,8	80,1	41,8	39,6	54,9	35,5	43,0	
AMU_779	Prp 19 beta protein	Prp 19	Q4ADG5	Mouse	57,3	6,1	30,6	166,0	107,1	99,1	109,4	99,2	103,7	123,5	133,0	137,8	118,7	128,2	

Figure 13.1: Overview of the differentially regulated proteins of the samples E (5 % CO_2) and F (13.3 % CO_2) with a t-test with $\alpha < 0.01$.

Bisher erschienendde Bände der Reihe

Bielefelder Schriften zur Zellkulturtechnik

ISSN: 1866-9727

1	Marc Wingens	Differentielle Proteomanalyse in der Zellkulturtechnik - Methodenentwicklung und Anwendung	
		ISBN 978-3-8325-1938-4	40.50 €
2	Michael Schomberg	Untersuchung von Glykosylierungsleistung und intrazellulären Nukleotidzucker-Konzentrationen rekombinanter CHO-Zellen unter hypothermen Kulturbedingungen	
		ISBN 978-3-8325-2448-7	46.00 €
3	Benedikt Greulich	Development of the CEMAX system for cell line development based on site-specific integration of expression cassettes	
		ISBN 978-3-8325-3112-6	39.00 €
4	Betina da Silva Ribeiro	Metabolic and bioprocess engineering of production cell lines for recombinant protein production	
		ISBN 978-3-8325-3123-2	53.50 €
5	Martin Heitmann	Vergleichende Untersuchung des Wachstums und des Metabolismus von Chinese Hamster Ovary Zellen in Kulturen bei niedrigen und hohen Zelldichten	
		ISBN 978-3-8325-3195-9	38.00 €
6	Tobias Thüte	Untersuchung der Hyperproduktivität tierischer Zellkulturen mittels *metabolomics*-Techniken als Tool der funktionellen Genomanalyse	
		ISBN 978-3-8325-3210-9	44.50 €
7	Sebastian Scholz	Integrierte Metabolom- und Transkriptomanalysen der humanen AGE1.HN Zelllinie. Eine Betrachtung aus systembiologischer Sicht	
		ISBN 978-3-8325-3301-4	58.50 €
8	Tim Frederik Beckmann	Differentielle Proteomanalyse tierischer Zellkulturen zur Charakterisierung von Hochproduzenten	
		ISBN 978-3-8325-3378-6	44.00 €

9	Benjamin Müller	Differentielle Analyse des Phosphoproteoms Apoptose-induzierter Jurkat ACC 282-Zellen
		ISBN 978-3-8325-3476-9 — 38.00 €
10	Sandra Klausing	Optimierung von CHO Produktionszelllinien: RNAi-vermittelter Gen-*knockdown* und Untersuchungen zur Klonstabilität
		ISBN 978-3-8325-3594-0 — 55.40 €

Alle erschienenen Bücher können unter der angegebenen ISBN im Buchhandel oder direkt beim Logos Verlag Berlin (www.logos-verlag.de, Fax: 030 42 85 10 92) bestellt werden.